Addition Puzzles for Kids

30 Puzzles Adding from 0-20

for kids ages 6-9

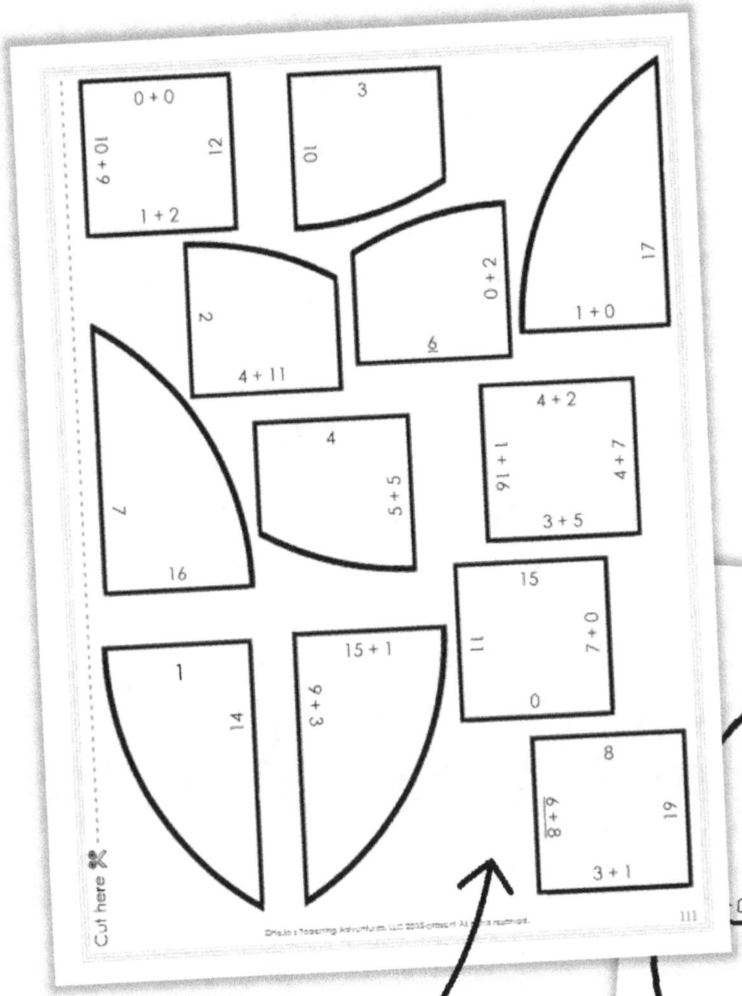

The goal is simple - practice addition facts by cutting out the puzzle pieces and putting them back together by answering the addition problems. Once complete, you'll reveal a fun image!

Go from this...

...to this!

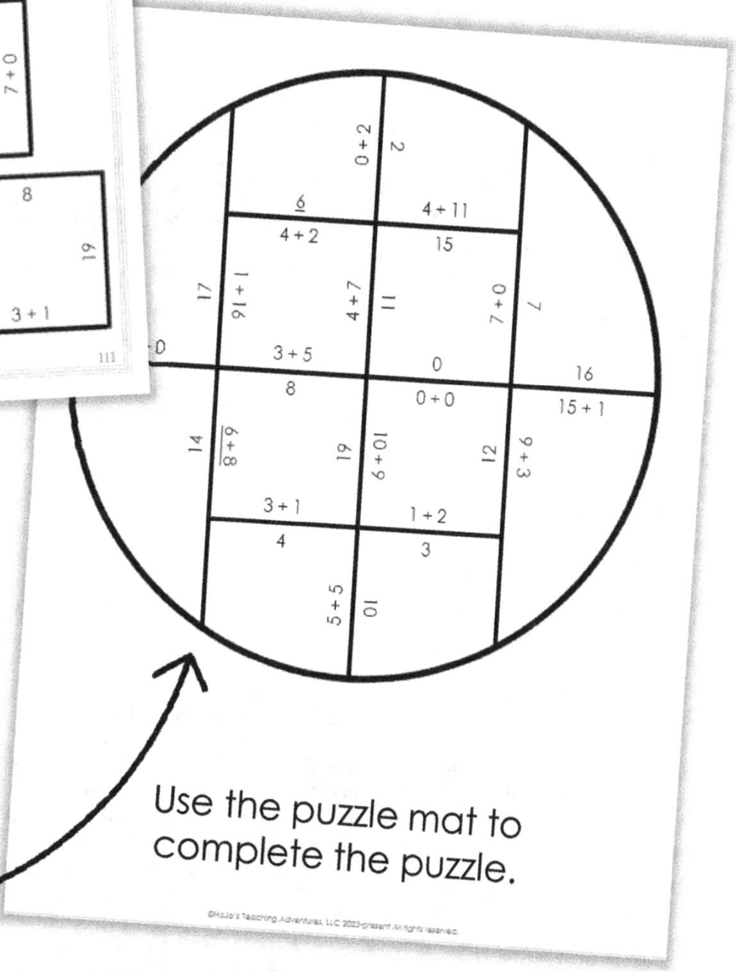

Contact the author :
www.HoJosTeachingAdventures.com

Table of Contents	Page

Instructions for Use

These puzzles were created to keep the practice of addition facts **hands-on**, **engaging**, and **FUN**. Help kids work on and memorize their addition facts within 20. Choose the puzzle you want, cut on the dotted line, and hand it to the child. The child will cut out the individual puzzle pieces and then put the puzzle back together.

If you want a bit more support for the child, also give them the puzzle mat so they can see the shape they are working on. These puzzle mats are optional, but they do offer more support for kids who are new to these puzzles or those who need a bit more help as they work through the addition facts.

There are **30 different puzzle shapes** to choose from, so you're sure to find something that will work for ANY time of year!

Here are the puzzle options included: snowman, mitten, penguin, mug, groundhog, heart, clover, horseshoe, bunny, egg, bee, raindrop, flip flop, apple, leaf, acorn, pumpkin, candy corn, cat, turkey, football, sneaker, gingerbread man, candy cane, stocking, tooth, circle, square, triangle, and star.

Ideas for Extra Support:

- Use the included hundred chart or number line for students who may be struggling.
- Allow the use of a calculator. (The memorization will come, but some students need support for longer than others.)
- Make sure they use the included puzzle mat for each shape!

Extended Learning:

- Save the puzzles and let students complete them more than one time.
- Have students draw their own shape, make cross lines to create puzzles like the ones in this book, and then write out their own addition problems. Cut and you'll have another puzzle to put together.

Answer keys for all 30 puzzles have been included at the end of this book.

I hope you enjoy these puzzles as much as I have enjoyed creating them! This teacher turned mom THANKS YOU for your interest in my work! Happy learning!!

Cut here

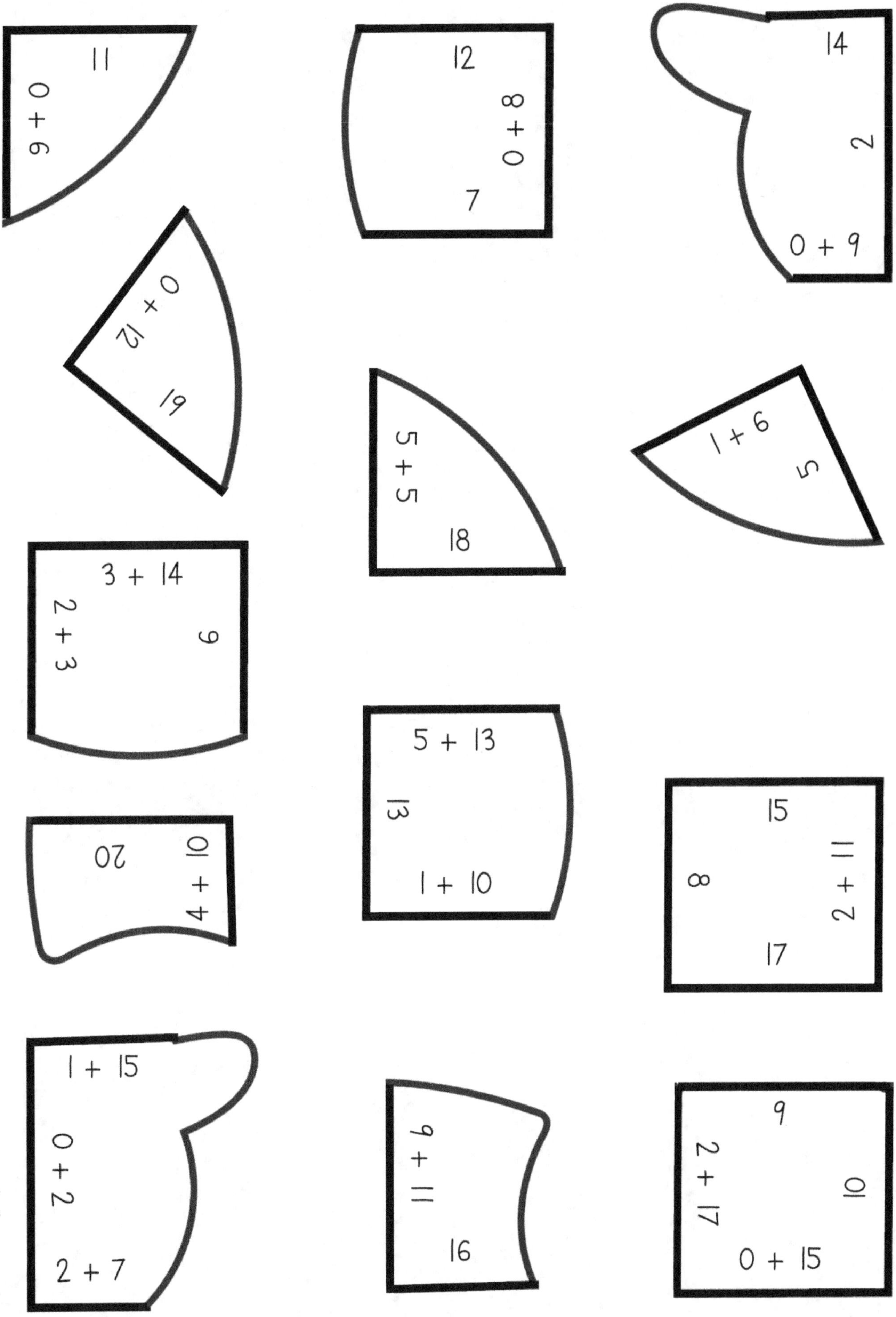
11
0 + 9
12
8 + 0
7
14
2
0 + 9
0 + 12
19
5 + 5
18
1 + 6
5
3 + 14
2 + 3
9
5 + 13
13
1 + 10
20
4 + 10
15
8
2 + 11
17
1 + 15
0 + 2
2 + 7
9 + 11
16
9
2 + 17
10
0 + 15

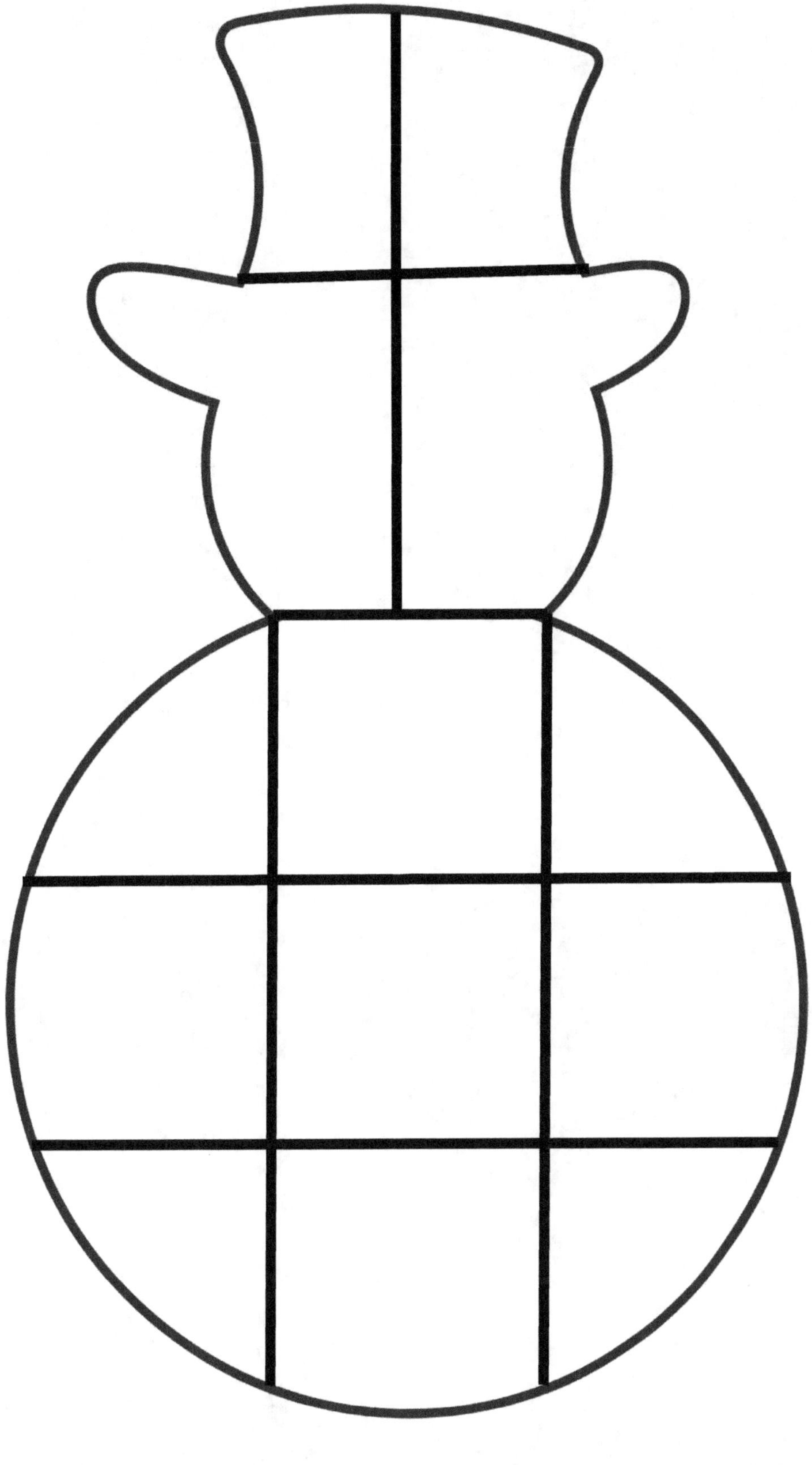

Use the puzzle mat to complete the puzzle.

Cut here ✂

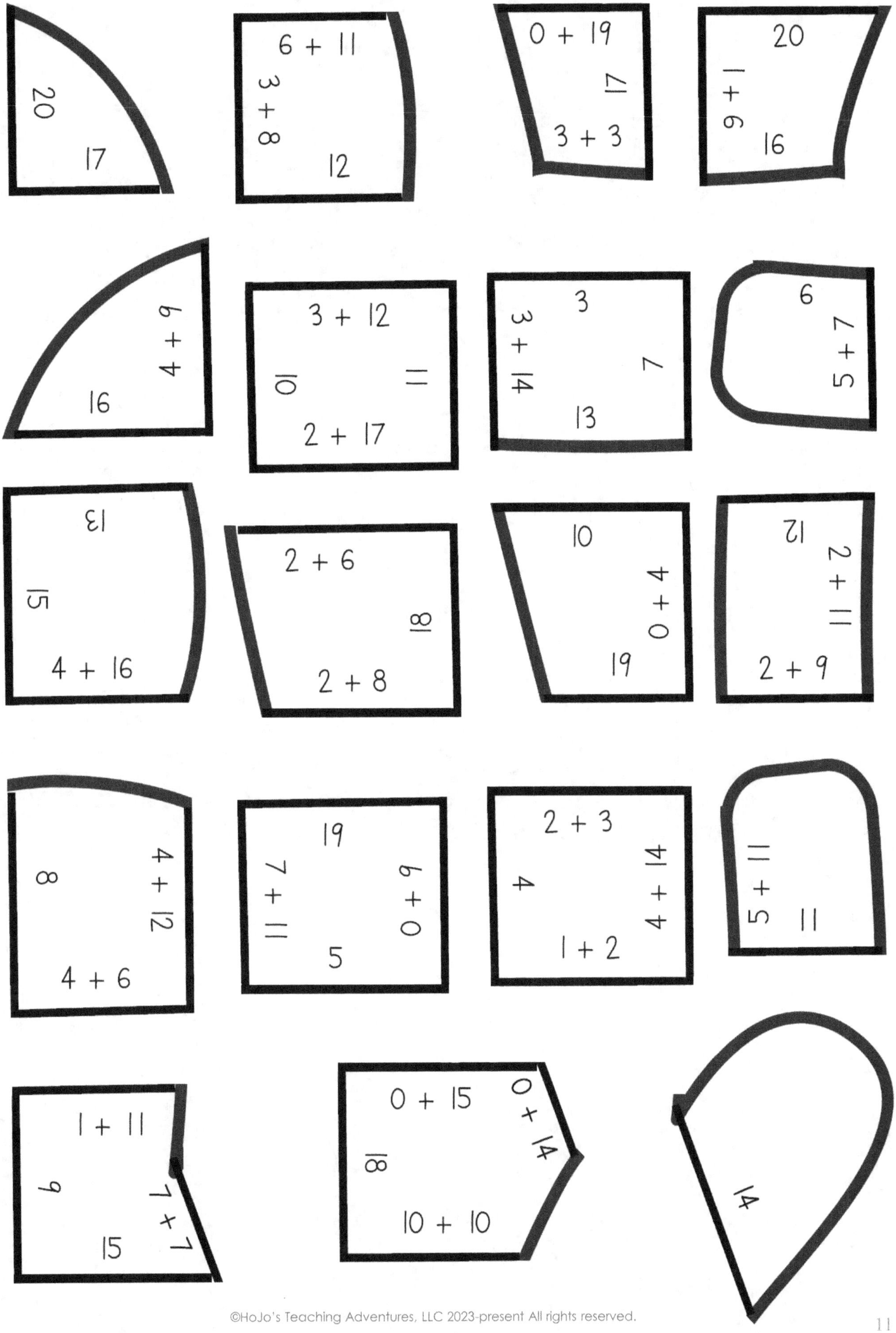
20
17
6 + 11
3 + 8
12
0 + 19
17
3 + 3
20
1 + 6
16
4 + 9
16
3 + 12
10
11
2 + 17
3
3 + 14
7
13
6
5 + 7
13
15
4 + 16
2 + 6
18
2 + 8
10
0 + 4
19
12
2 + 11
2 + 9
8
4 + 12
4 + 6
19
7 + 11
0 + 9
5
2 + 3
4
4 + 14
1 + 2
5 + 11
11
1 + 11
9
7 + 7
15
0 + 15
0 + 14
18
10 + 10
14

Use the puzzle mat to complete the puzzle.

Cut here

13
9
4
2 + 5
4 + 5
0 + 0
5
3 + 3
5 + 6
15
1 + 1
3
5 + 11
0 + 1
11
6 + 7
3 + 12
2 + 2
0 + 3
2 + 10
5 + 5
1 + 4
8
12
5 + 9
16
0
3 + 5
10
1
7
14
2
6

Use the puzzle mat to complete the puzzle.

Cut here ✂

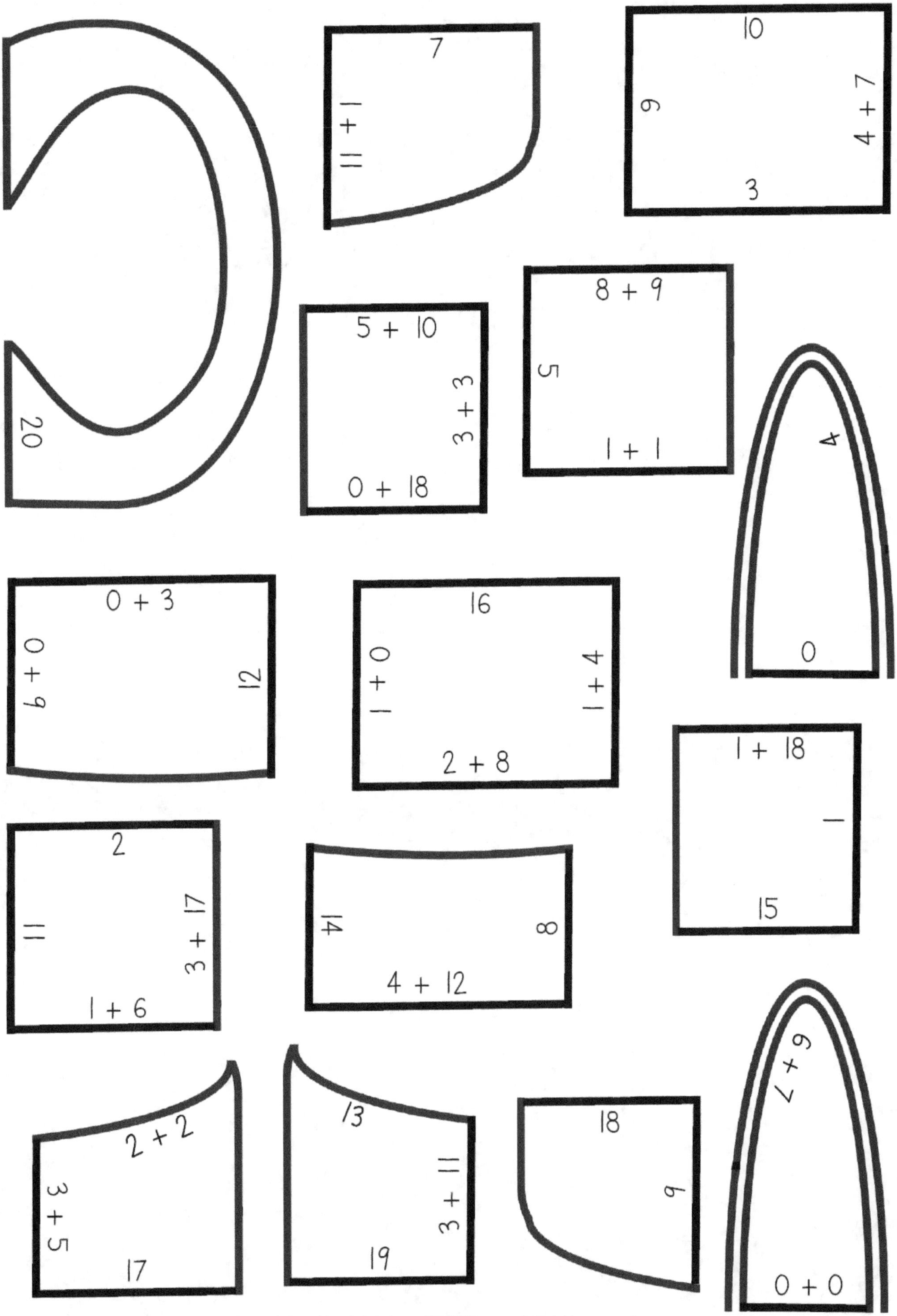
20
7
11 + 1
10
9
4 + 7
3
5 + 10
3 + 3
0 + 18
8 + 9
5
1 + 1
4
0
0 + 3
0 + 9
12
16
1 + 0
1 + 4
2 + 8
1 + 18
1
15
2
11
3 + 17
1 + 6
14
8
4 + 12
6 + 7
2 + 2
3 + 5
17
13
3 + 11
19
18
9
0 + 0

Use the puzzle mat to complete the puzzle.

Cut here

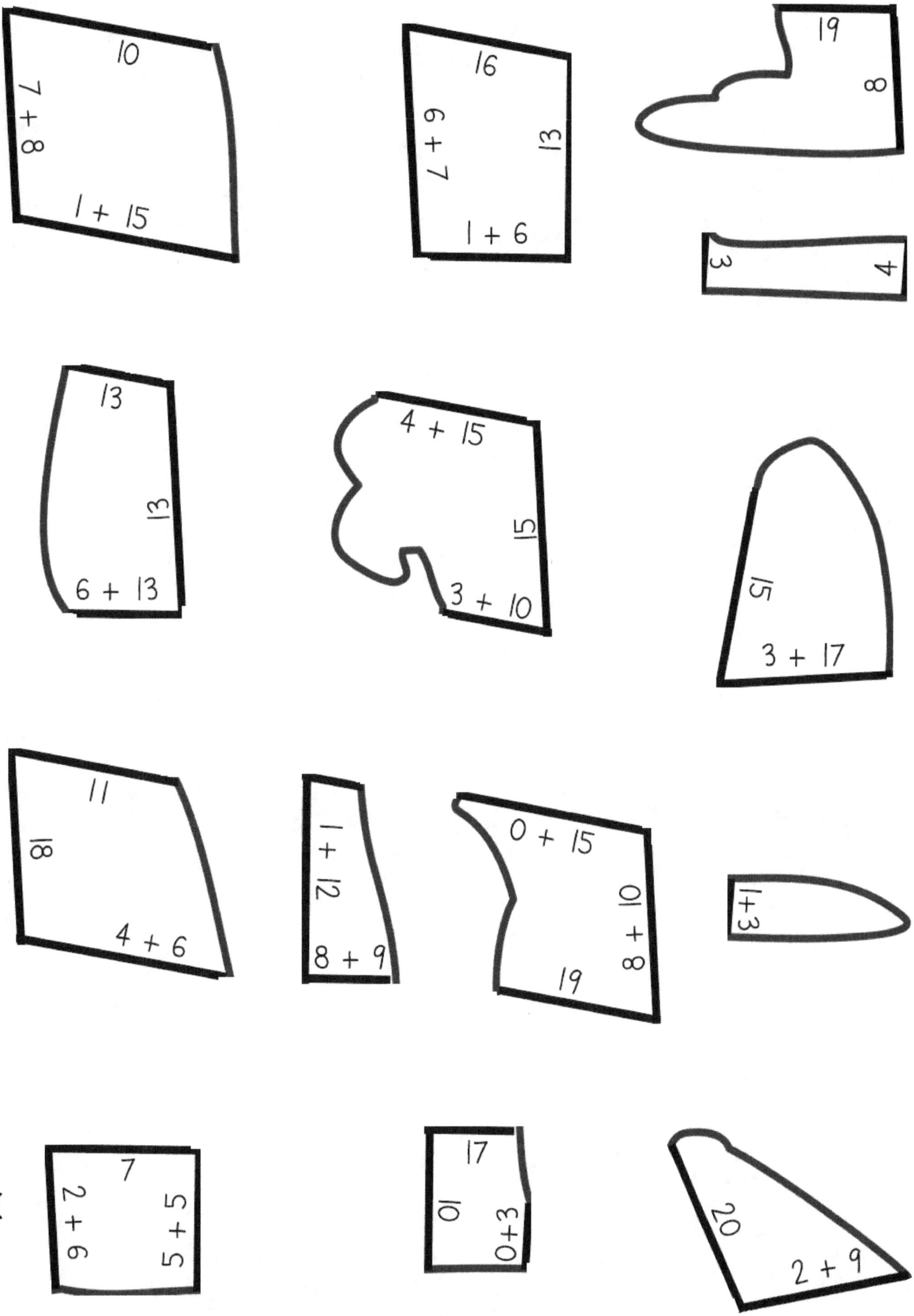
10
7 + 8
1 + 15
16
6 + 7
13
1 + 6
19
8
3
4
13
13
6 + 13
4 + 15
15
3 + 10
15
3 + 17
11
18
4 + 6
1 + 12
8 + 9
0 + 15
8 + 10
19
1+3
7
2 + 6
5 + 5
17
10
0+3
20
2 + 9

Use the puzzle mat to complete the puzzle.

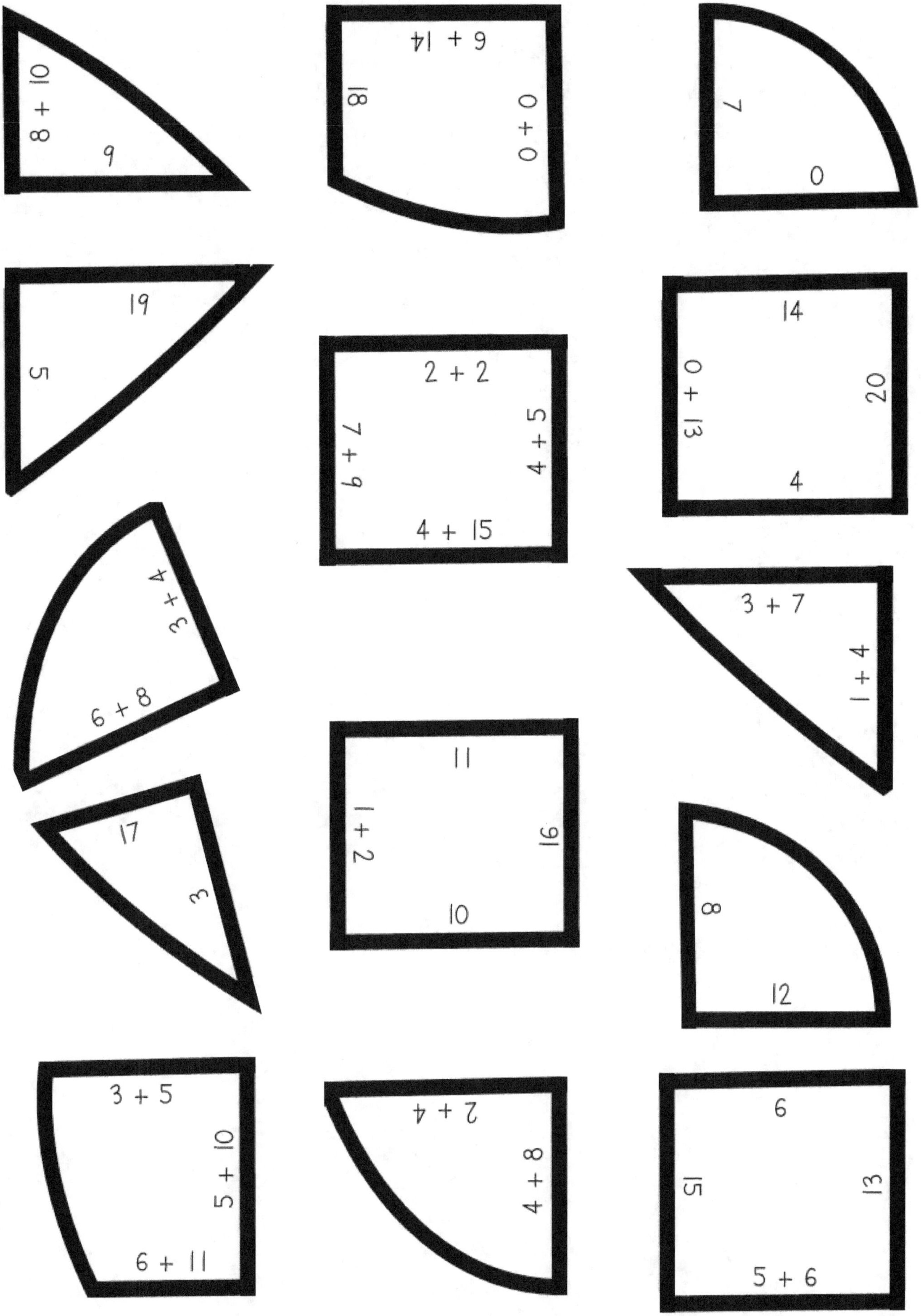
8 + 10
9
6 + 14
18
0 + 0
7
0
19
5
14
0 + 13
20
4
2 + 2
7 + 9
4 + 5
4 + 15
3 + 4
6 + 8
3 + 7
1 + 4
17
3
11
1 + 2
16
10
8
12
3 + 5
5 + 10
6 + 11
2 + 4
4 + 8
6
15
13
5 + 6

Cut here ✂

Use the puzzle mat to complete the puzzle.

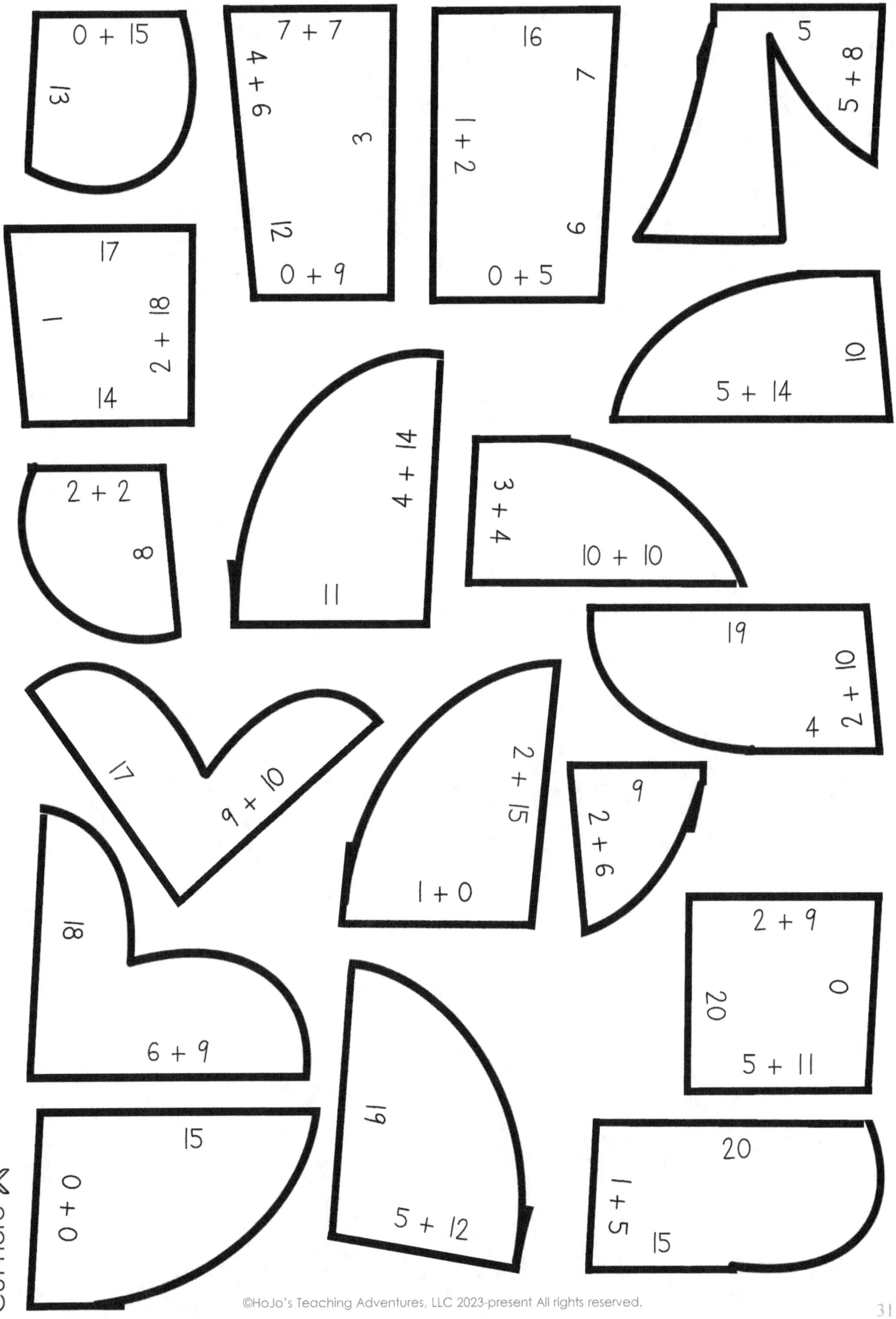
0 + 15
13
7 + 7
4 + 6
3
12
0 + 9
16
7
1 + 2
9
0 + 5
5
5 + 8
17
1
2 + 18
14
5 + 14
10
2 + 2
8
4 + 14
11
3 + 4
10 + 10
19
2 + 10
4
17
9 + 10
2 + 15
1 + 0
9
2 + 6
18
6 + 9
2 + 9
20
0
5 + 11
15
0 + 0
19
5 + 12
20
1 + 5
15

Cut here ✂

Use the puzzle mat to complete the puzzle.

Cut here ✂

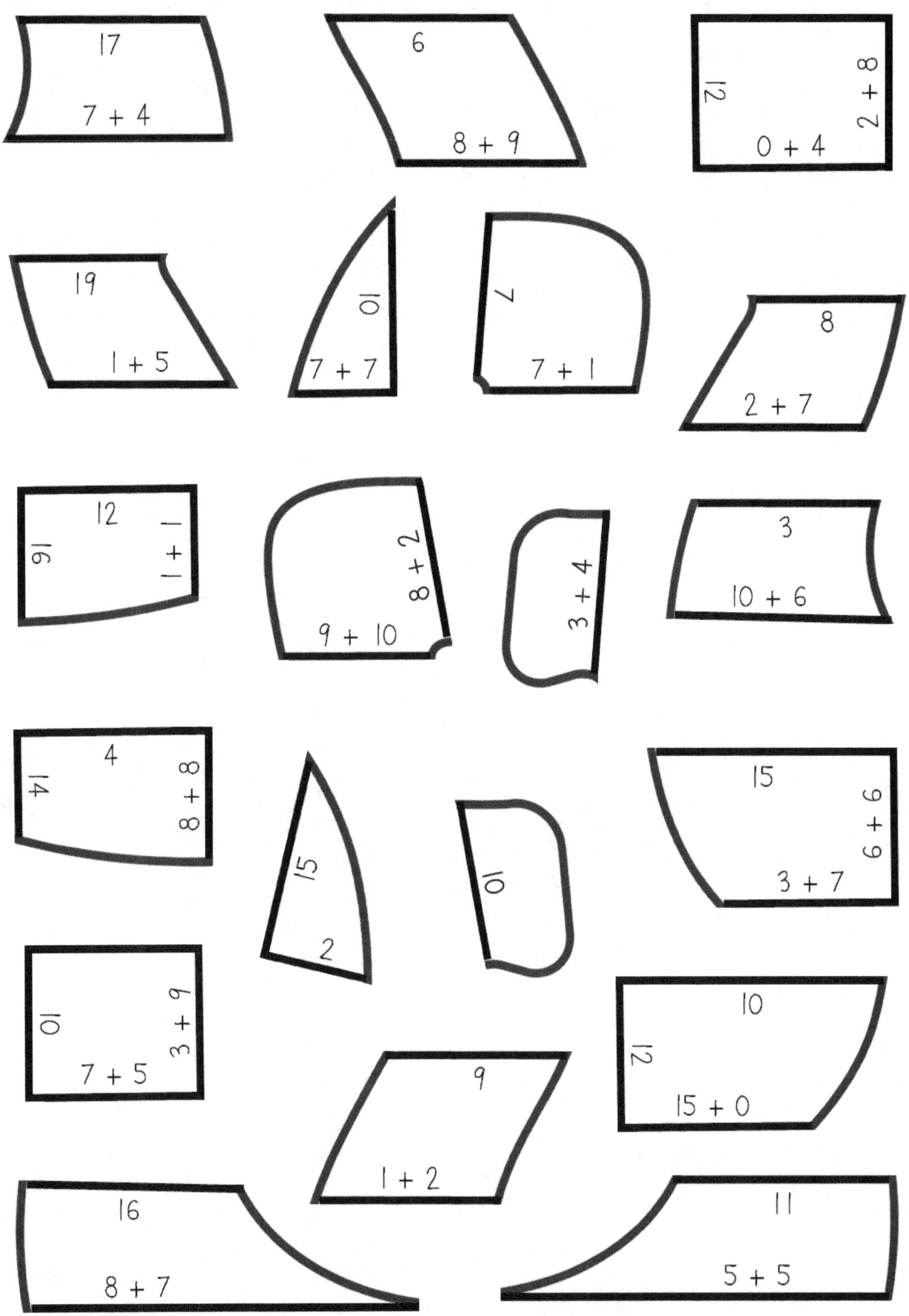
17
7 + 4
6
8 + 9
12
2 + 8
0 + 4
19
1 + 5
10
7 + 7
7
7 + 1
8
2 + 7
12
16
1 + 1
8 + 2
9 + 10
3 + 4
3
10 + 6
4
14
8 + 8
15
2
10
15
6 + 6
3 + 7
10
3 + 9
7 + 5
9
1 + 2
10
12
15 + 0
16
8 + 7
11
5 + 5

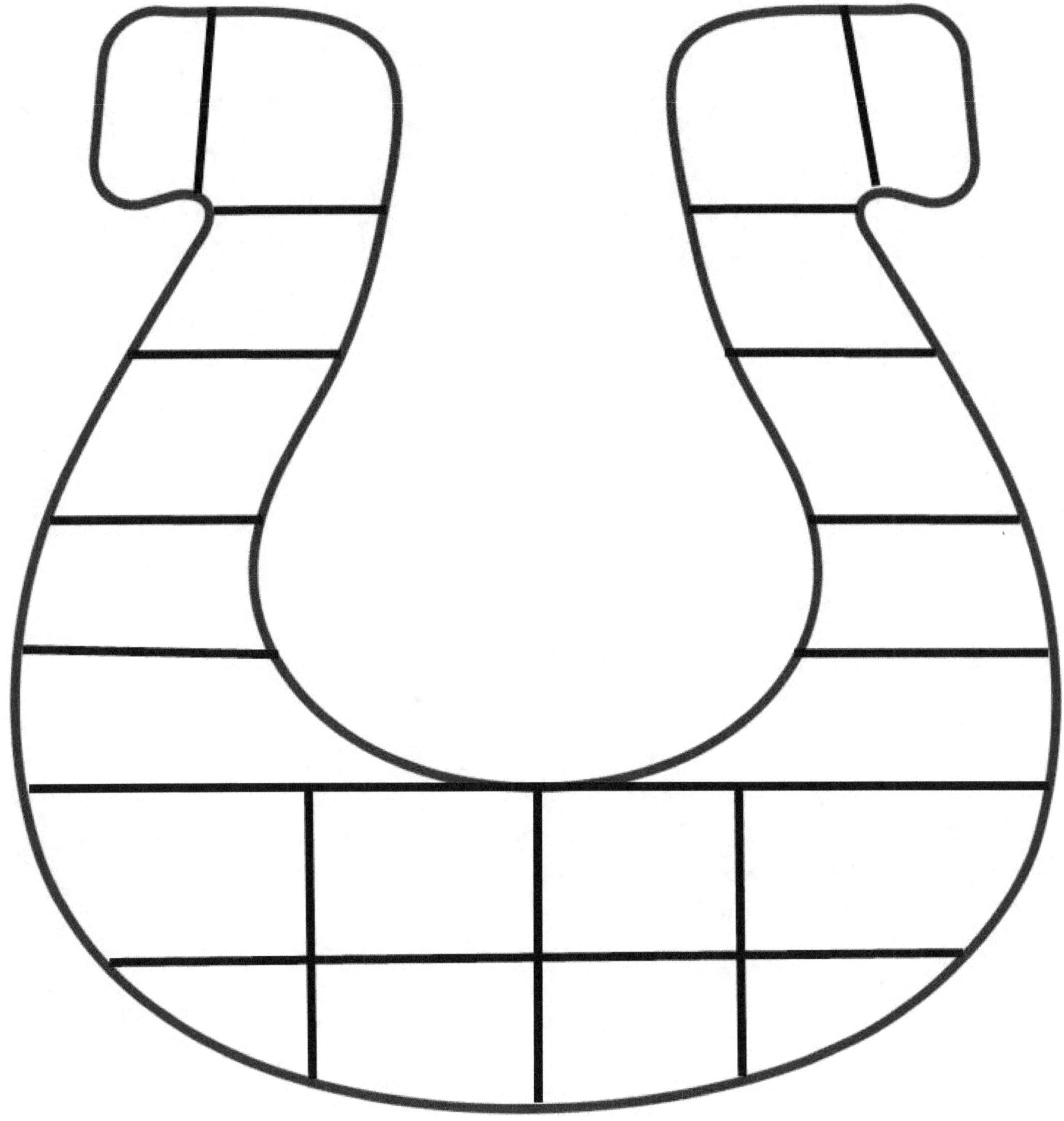

Use the puzzle mat to complete the puzzle.

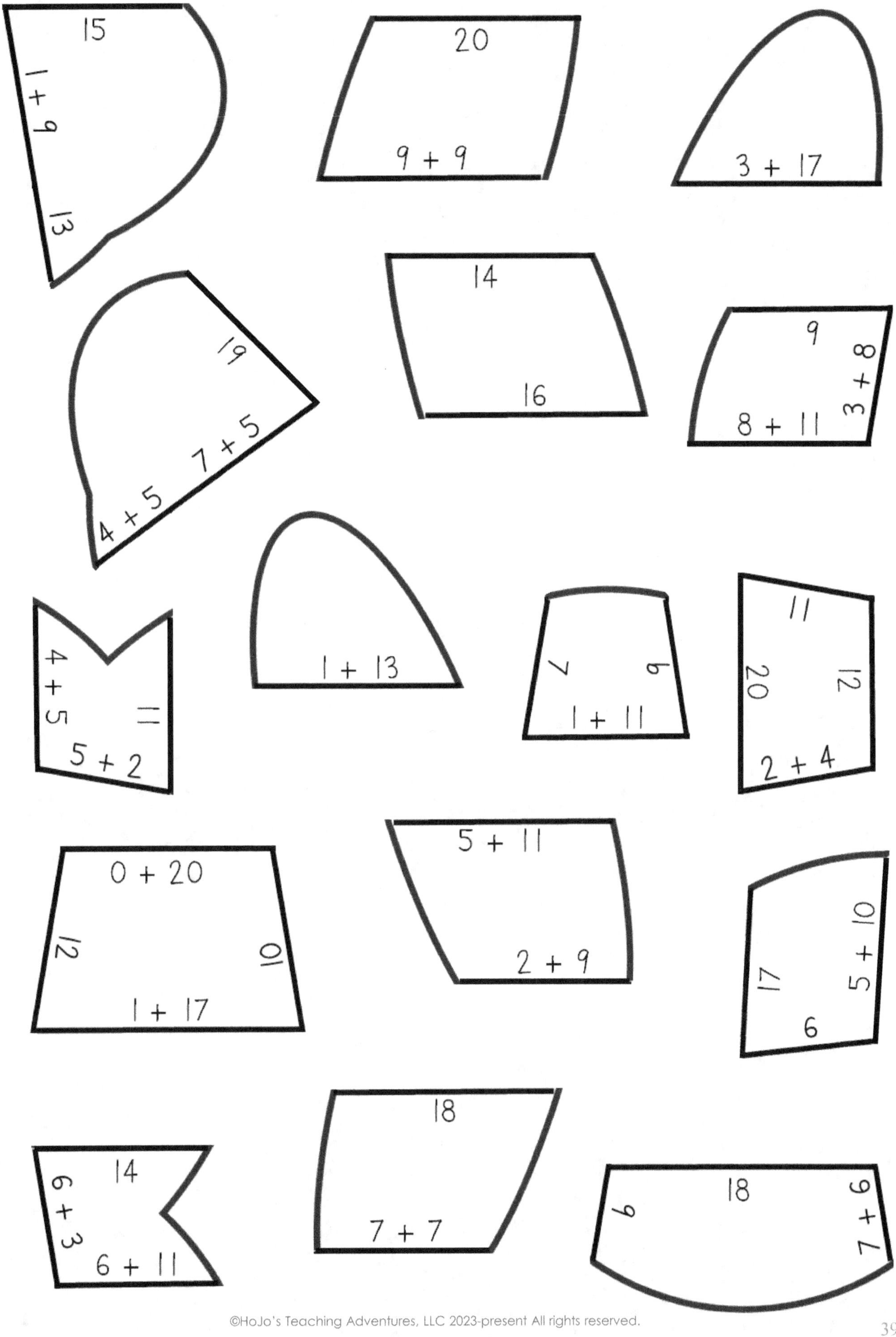
15
1 + 9
13
20
9 + 9
3 + 17
19
7 + 5
4 + 5
14
16
9
3 + 8
8 + 11
4 + 5
11
5 + 2
1 + 13
7
6
1 + 11
11
20
12
2 + 4
0 + 20
12
10
1 + 17
5 + 11
2 + 9
5 + 10
17
6
14
6 + 3
6 + 11
18
7 + 7
18
9
9 + 7

Cut here ✂

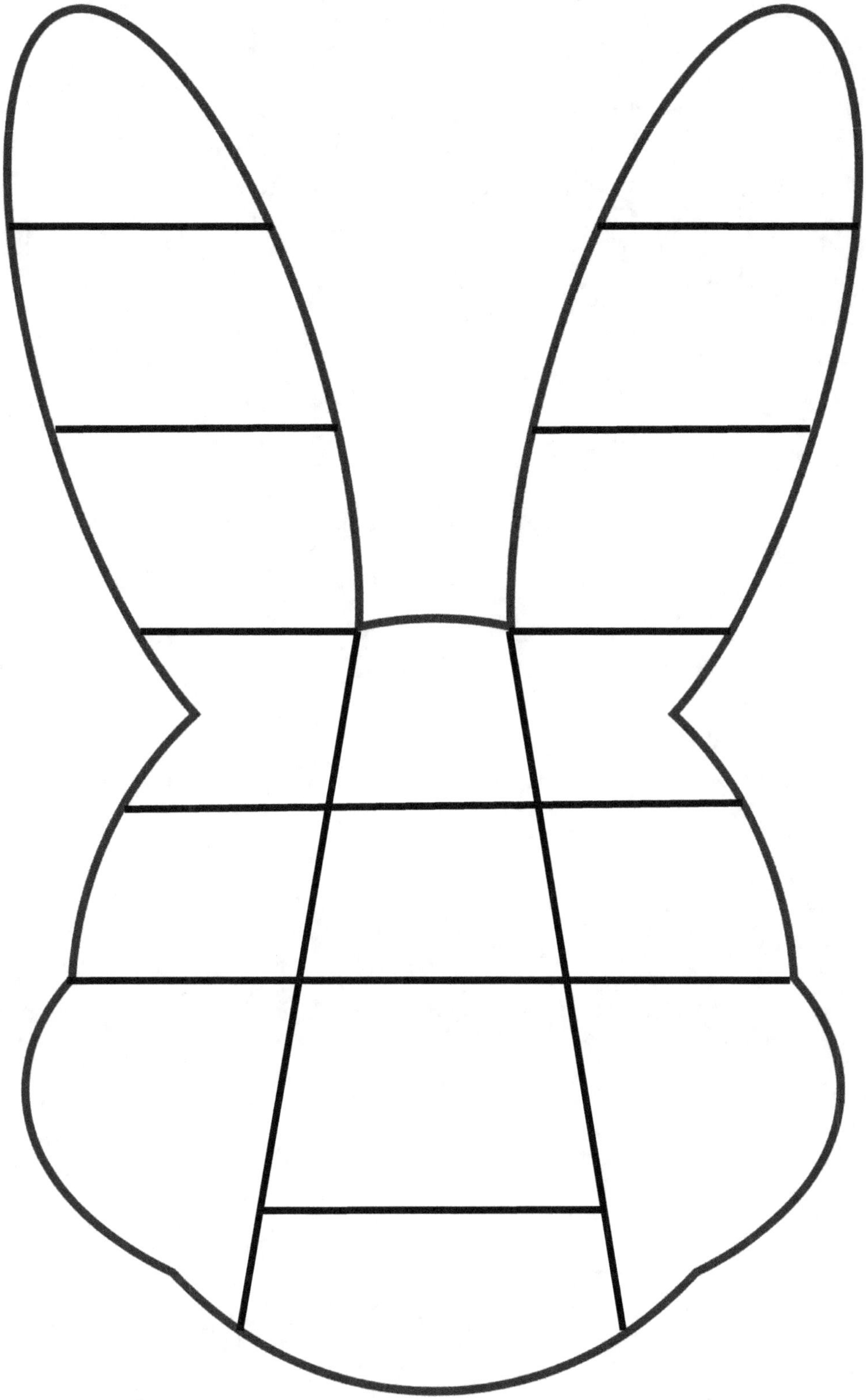

Use the puzzle mat to complete the puzzle.

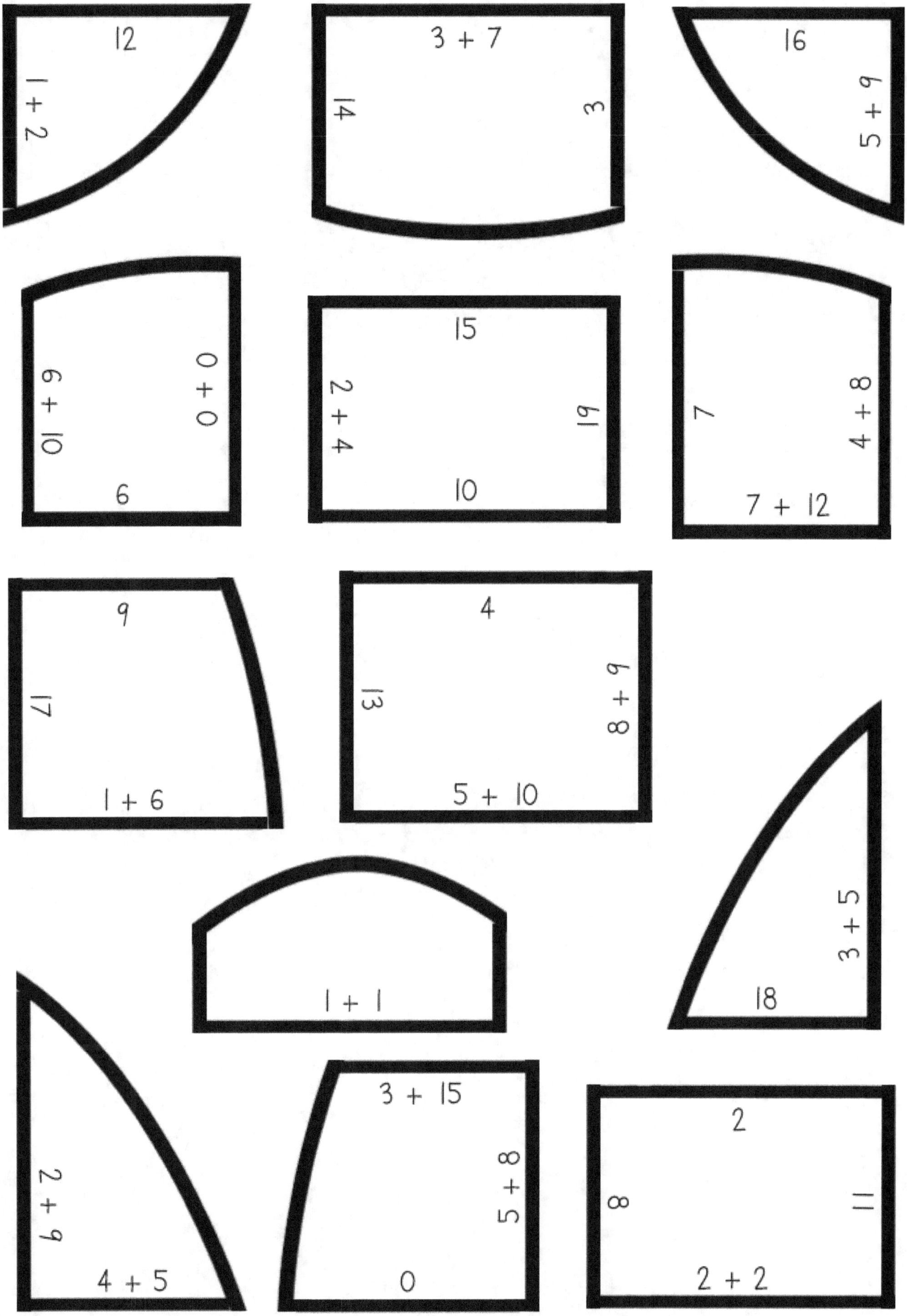
12
1 + 2
3 + 7
14
3
16
6 + 9
6 + 10
0 + 0
6
15
2 + 4
19
10
7
4 + 8
7 + 12
9
17
1 + 6
4
13
8 + 9
5 + 10
1 + 1
3 + 5
18
2 + 7
6 + 9
4 + 5
3 + 15
5 + 8
0
2
8
11
2 + 2

Cut here

Use the puzzle mat to complete the puzzle.

1 + 1
2 + 3
11 + 9
15
4 + 16
6
2 + 5
4
14
3 + 13
5
3 + 6
20
1 + 2
1 + 10
4 + 10
4 + 8
3 + 3
13
8
12
2
11
0
2 + 11
19
3
5 + 10
1
18
10
1 + 0
0 + 0
4 + 4
17
3 + 7
9
1 + 3
5 + 13
5 + 14
16
7

Cut here ✂

Use the puzzle mat to complete the puzzle.

Cut here

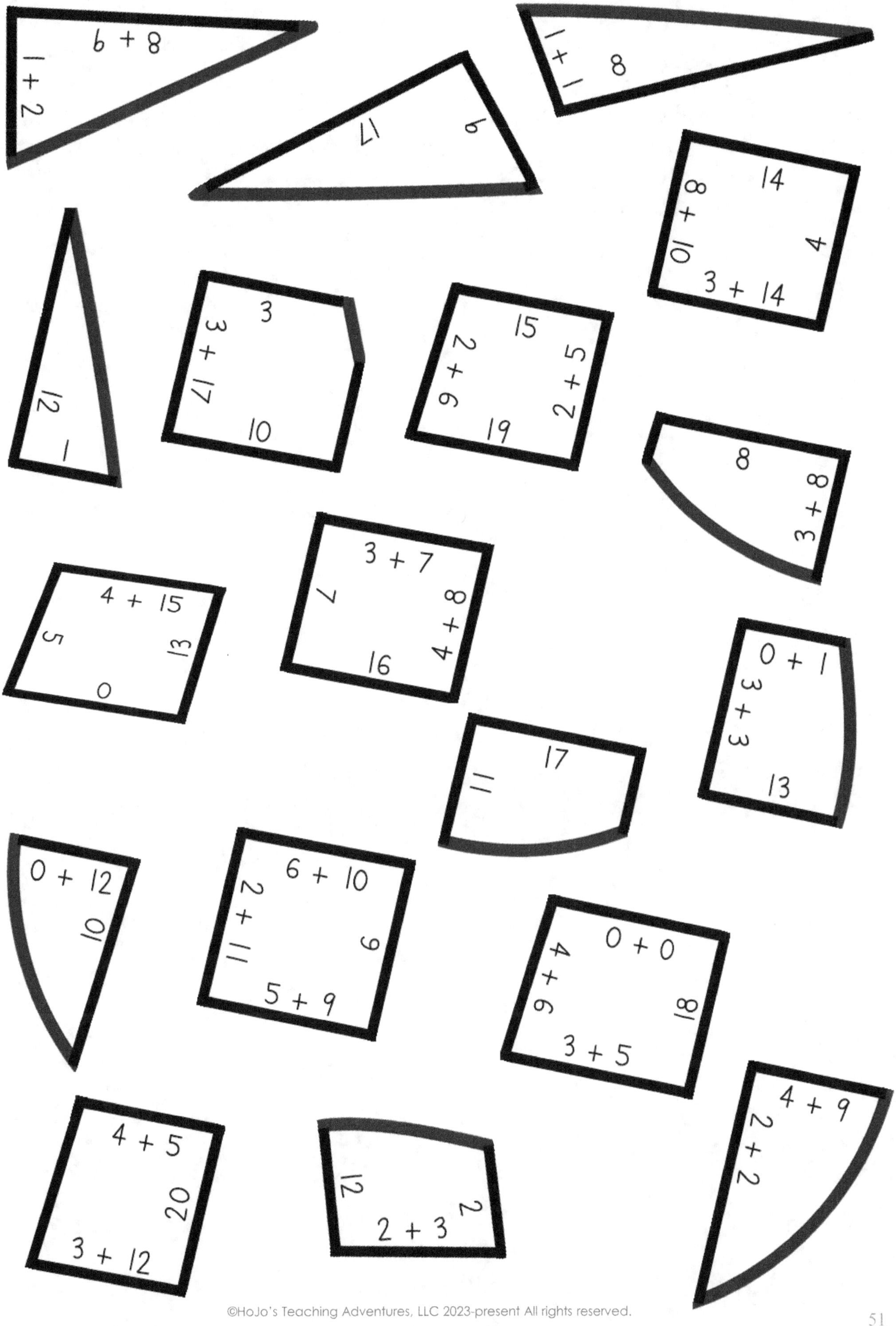
8 + 9
1 + 2
1 + 1
8
17
9
14
8 + 10
4
3 + 14
12
1
3
3 + 17
10
15
2 + 6
2 + 5
19
8
3 + 8
3 + 7
7
4 + 8
16
4 + 15
5
13
0
0 + 1
3 + 3
13
17
11
0 + 12
10
6 + 10
2 + 11
9
5 + 9
0 + 0
4 + 6
18
3 + 5
4 + 5
20
3 + 12
12
2
2 + 3
4 + 9
2 + 2

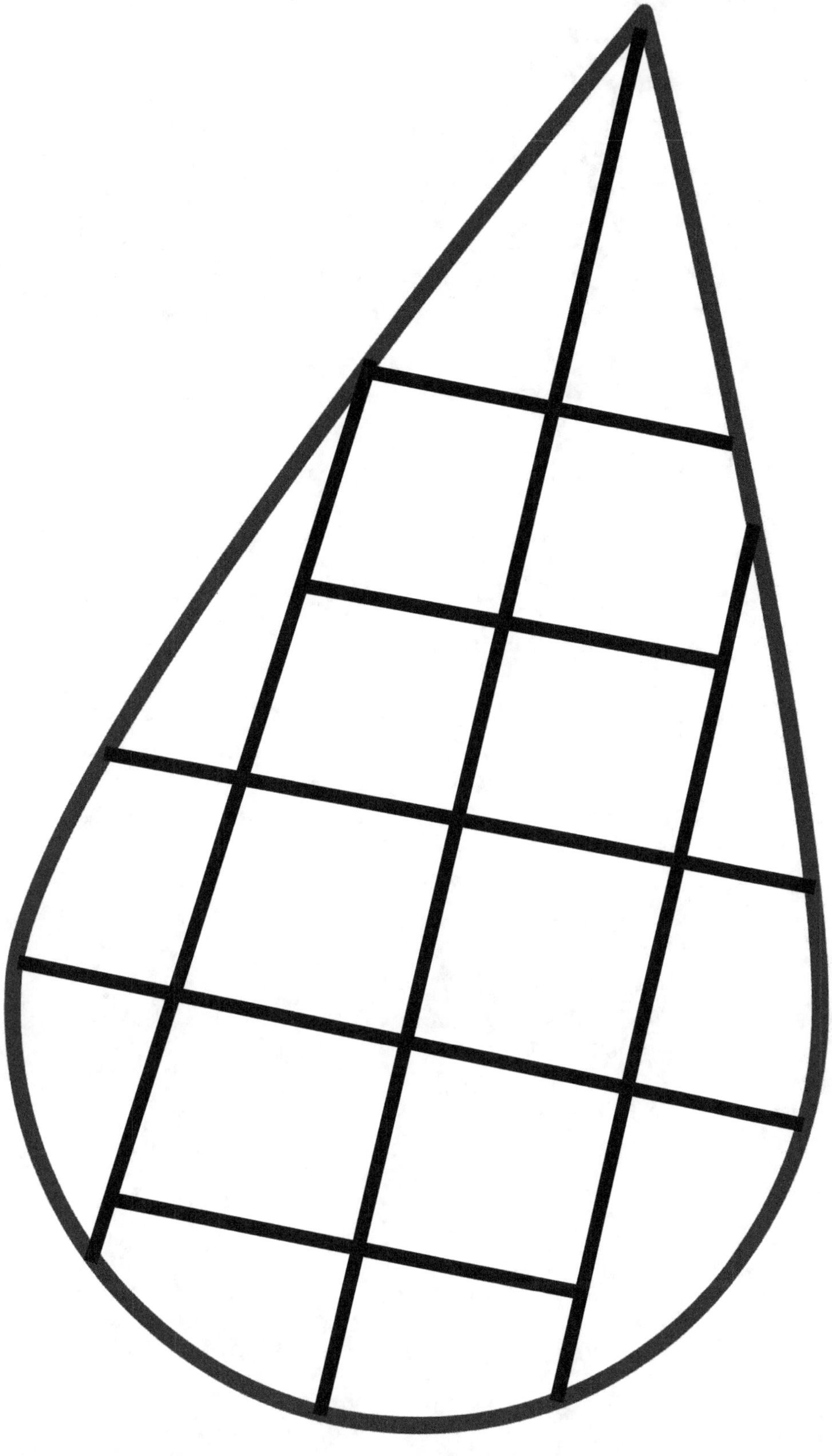

Use the puzzle mat to complete the puzzle.

3
20
0 + 15
0
10 + 10
7
17
11
2 + 2
4
3 + 4
2 + 3
5 + 12
20 + 0
5
9
4 + 5
5 + 6
20
3 + 7
6
0 + 0
13
1 + 5
15
7 + 11
18
1 + 2
10
6 + 7

Cut here

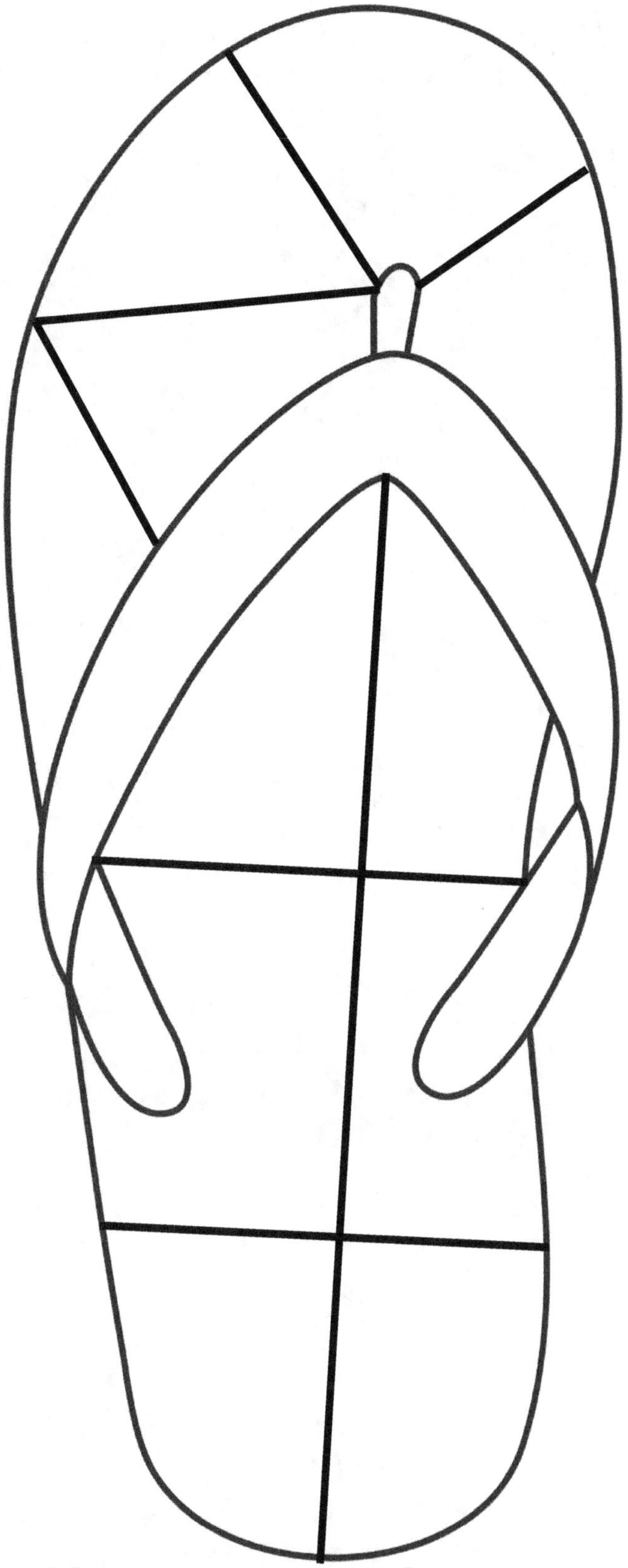

Use the puzzle mat to complete the puzzle.

Cut here ✂

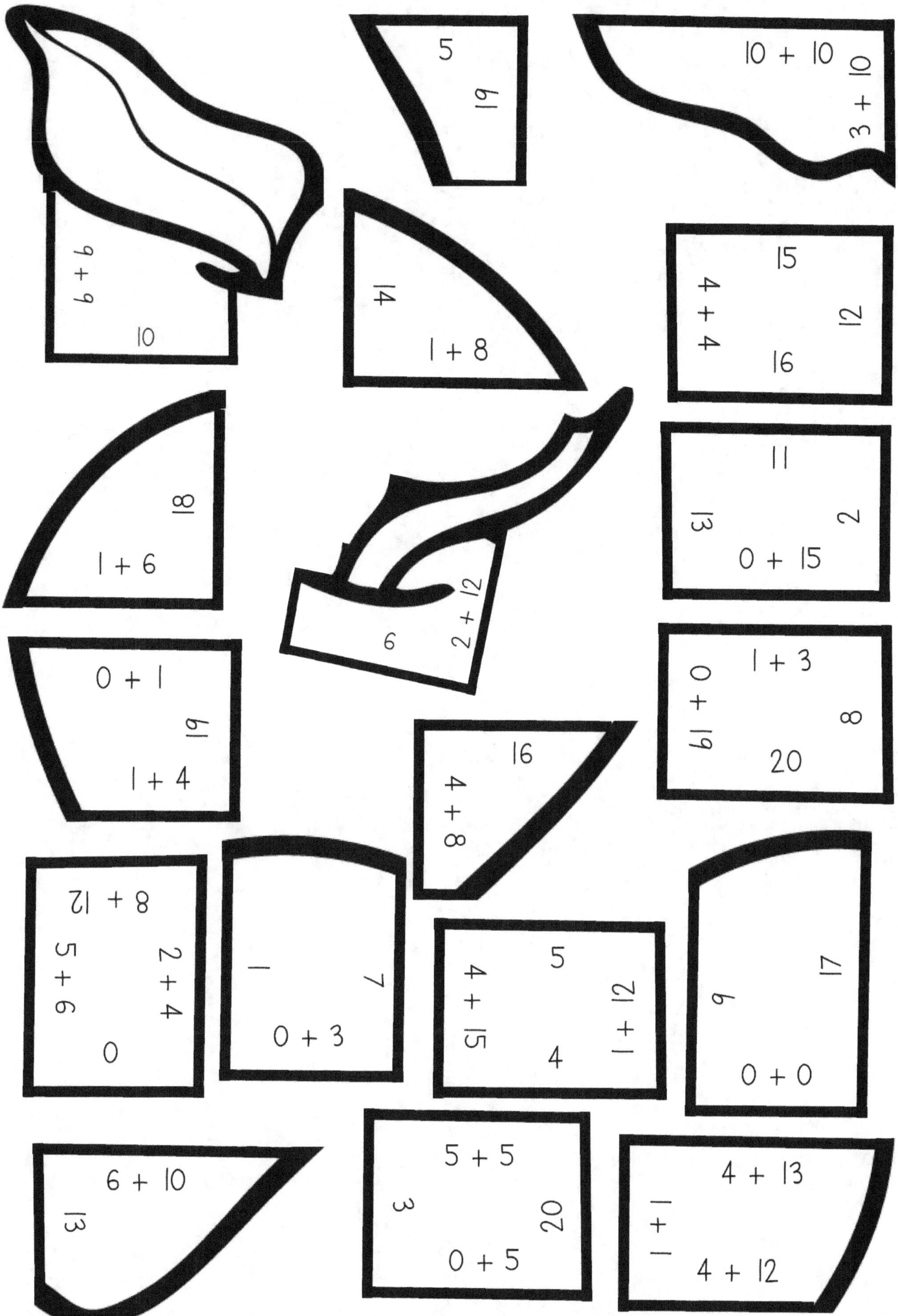
5
19
10 + 10
3 + 10
9 + 9
10
14
1 + 8
15
4 + 4
12
16
18
1 + 6
11
13
2
0 + 15
2 + 12
6
0 + 1
19
1 + 4
1 + 3
0 + 19
8
20
16
4 + 8
8 + 12
5 + 6
2 + 4
0
1
7
0 + 3
5
4 + 15
1 + 12
4
9
17
0 + 0
6 + 10
13
5 + 5
3
20
0 + 5
4 + 13
1 + 1
4 + 12

Use the puzzle mat to complete the puzzle.

Cut here

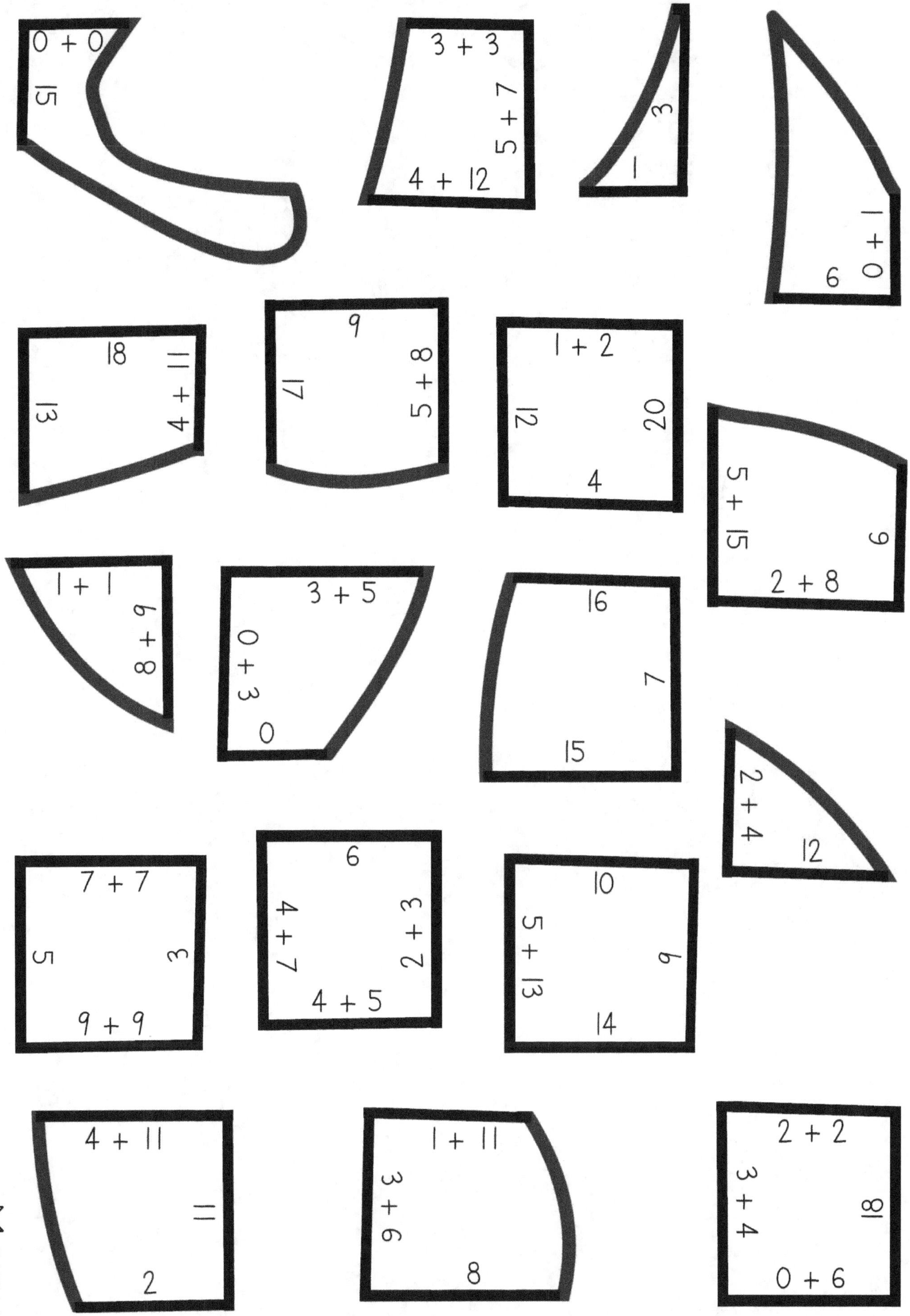
0 + 0
15
3 + 3
5 + 7
4 + 12
3
1
1 + 0
6
18
4 + 11
13
9
5 + 8
17
1 + 2
12
20
4
5 + 15
6
2 + 8
1 + 1
8 + 9
3 + 5
0 + 3
0
16
7
15
2 + 4
12
7 + 7
5
3
9 + 9
6
4 + 7
2 + 3
4 + 5
10
5 + 13
9
14
4 + 11
11
2
1 + 11
3 + 6
8
2 + 2
3 + 4
18
0 + 6

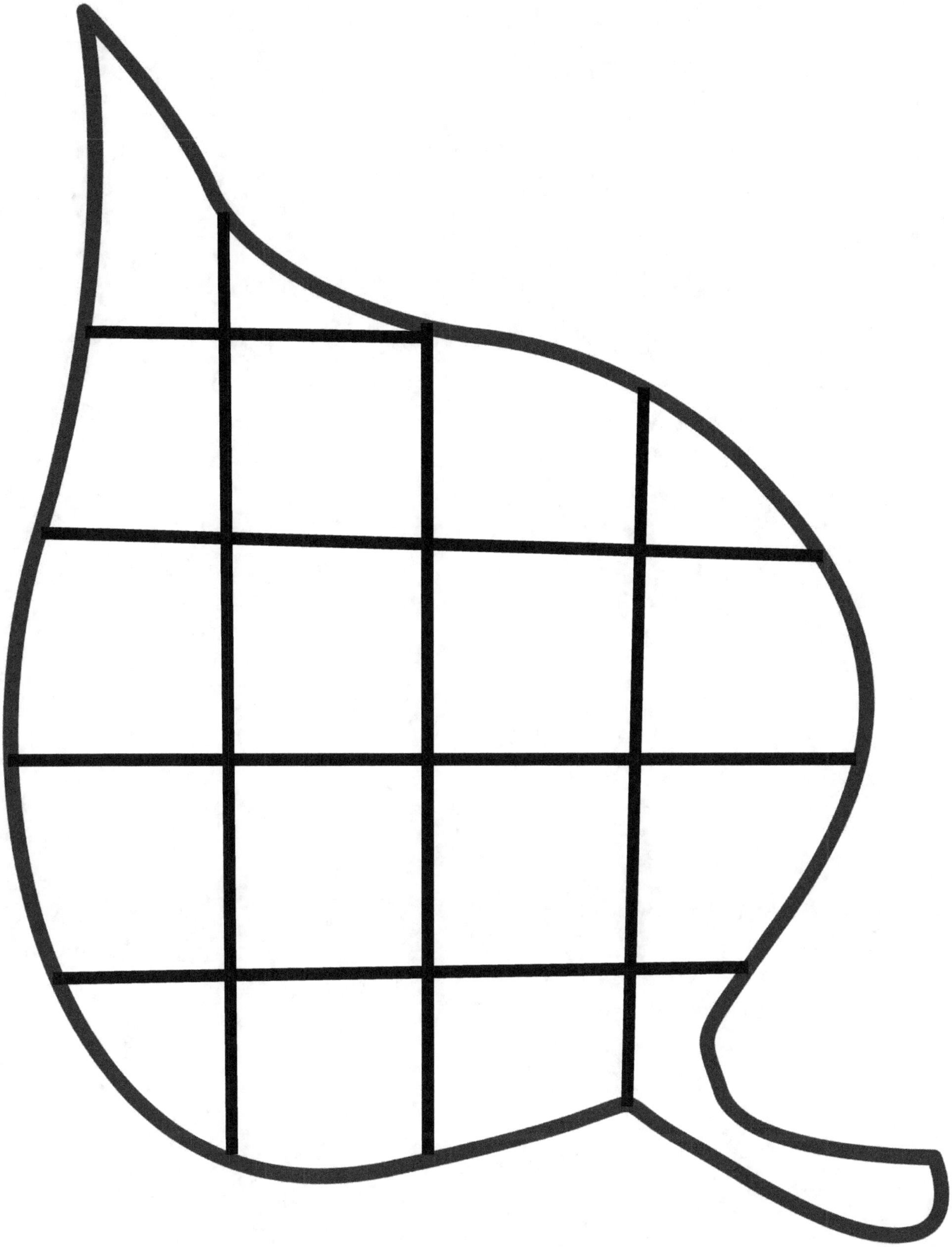

Use the puzzle mat to complete the puzzle.

Cut here

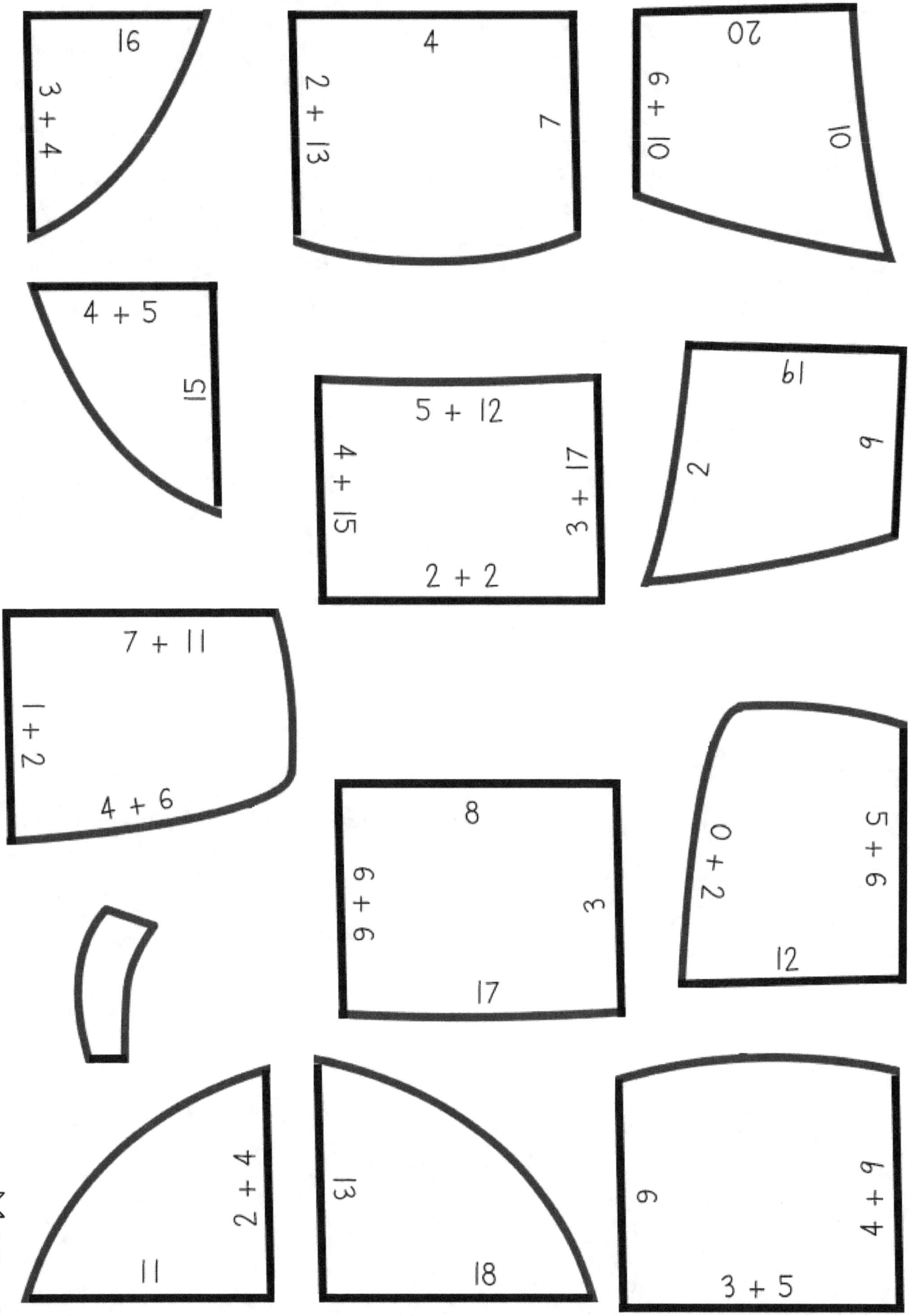

16
3 + 4
4
2 + 13
7
20
6 + 10
10
4 + 5
15
5 + 12
4 + 15
3 + 17
2 + 2
19
2
6
7 + 11
1 + 2
4 + 6
8
6 + 6
3
17
0 + 2
5 + 6
12
2 + 4
2 + 7
11
13
18
6
6 + 4
4 + 4
3 + 5

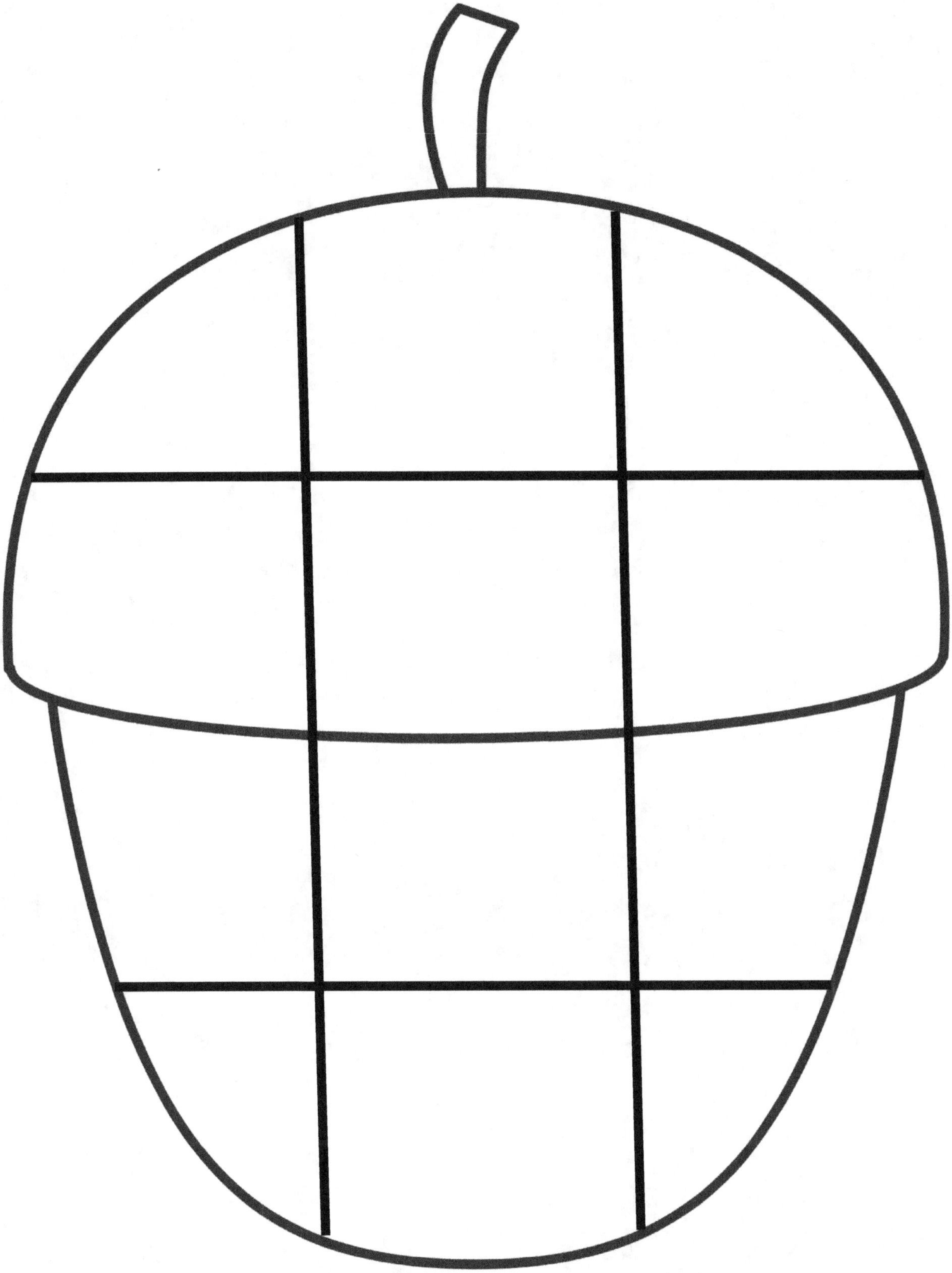

Use the puzzle mat to complete the puzzle.

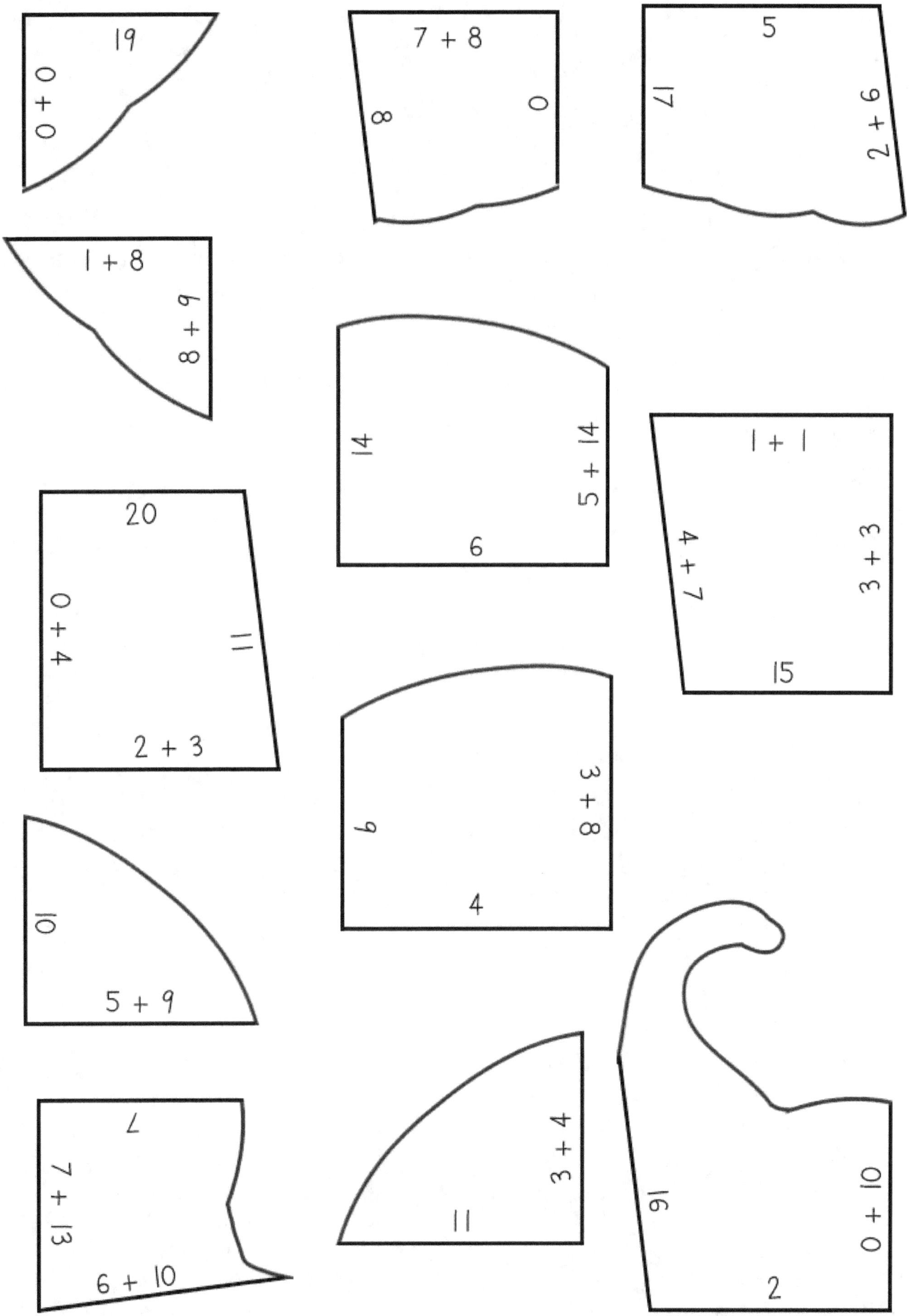
19
0 + 0
7 + 8
8
0
5
17
2 + 6
1 + 8
8 + 9
14
5 + 14
6
1 + 1
4 + 7
3 + 3
15
20
0 + 4
11
2 + 3
9
3 + 8
4
10
5 + 9
7
7 + 13
6 + 10
3 + 4
11
16
0 + 10
2

Cut here

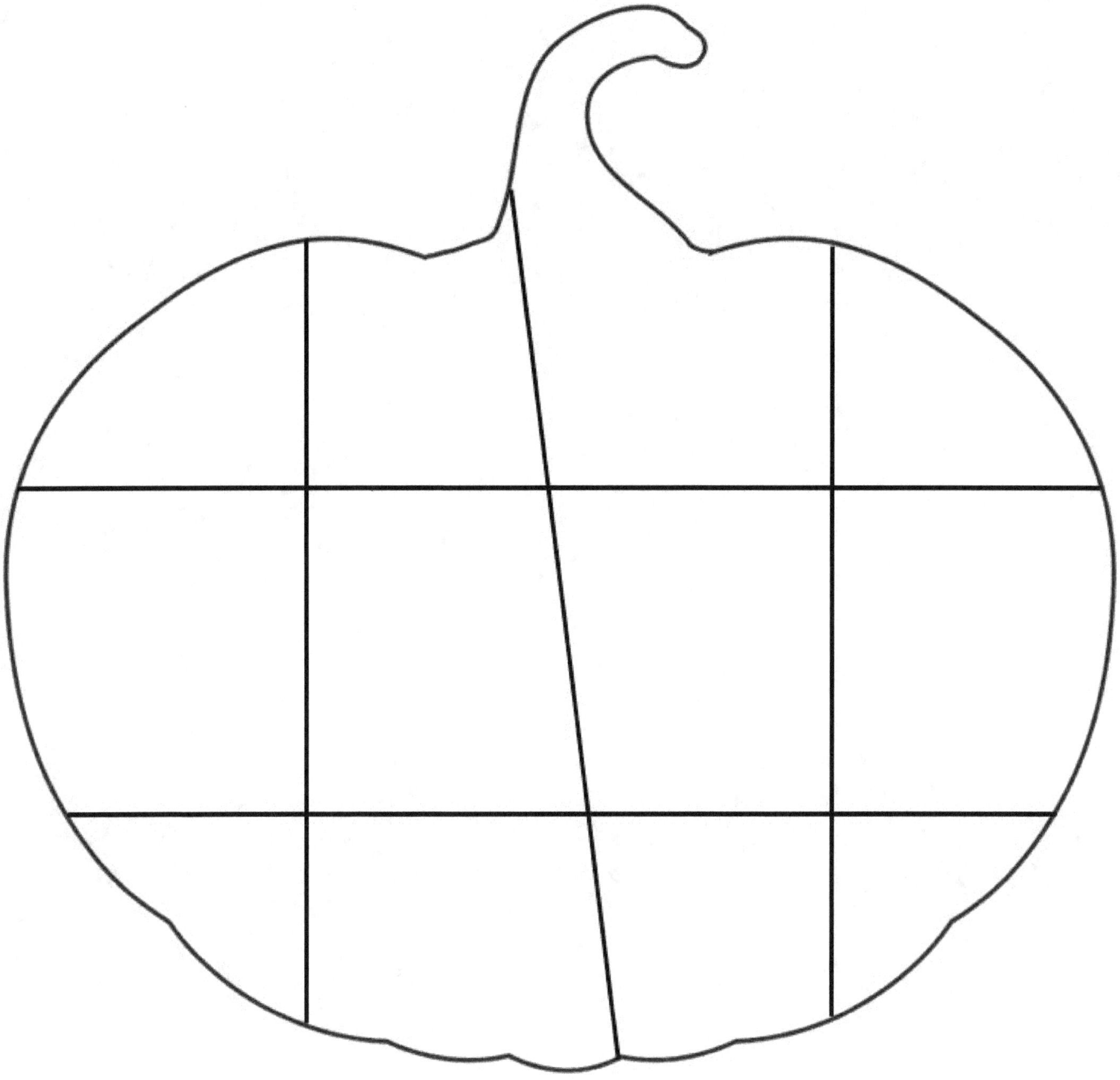

Use the puzzle mat to complete the puzzle.

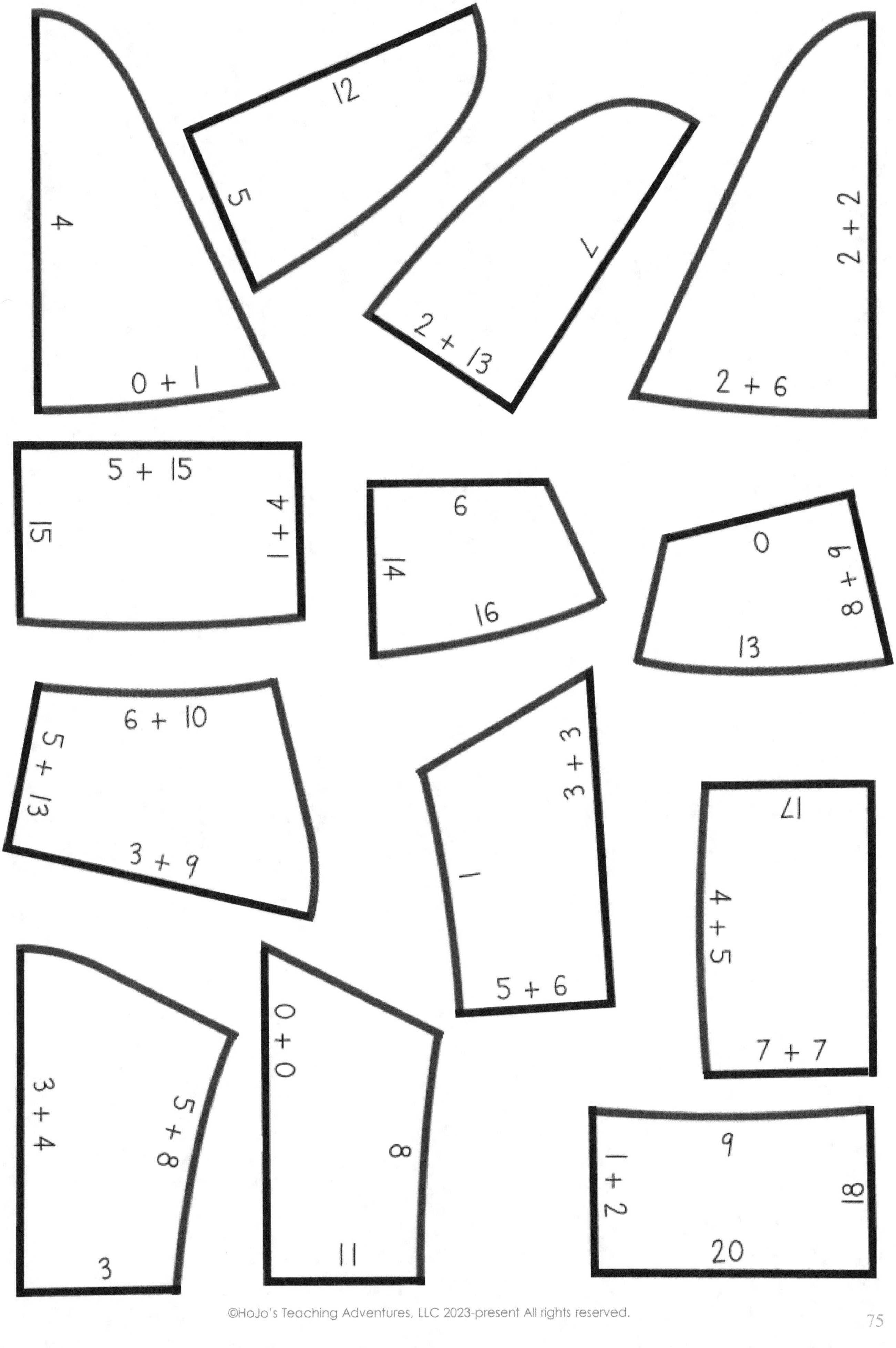
4
0 + 1
12
5
7
2 + 13
2 + 2
2 + 6
5 + 15
15
1 + 4
6
14
16
0
9 + 8
13
6 + 10
5 + 13
3 + 9
3 + 3
1
5 + 6
17
4 + 5
7 + 7
3 + 4
5 + 8
3
0 + 0
8
11
9
1 + 2
18
20

Cut here ✂

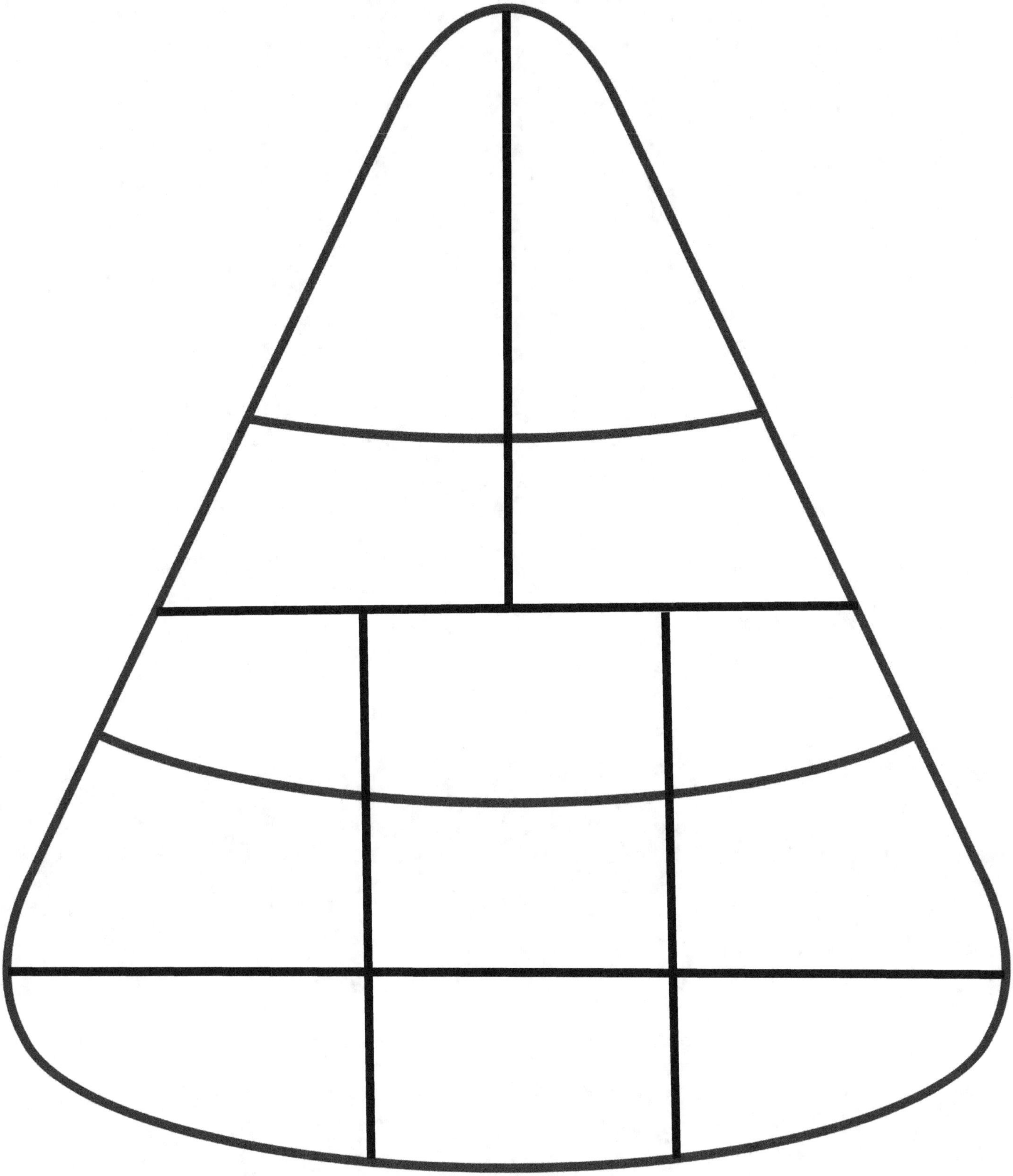

Use the puzzle mat to complete the puzzle.

Cut here

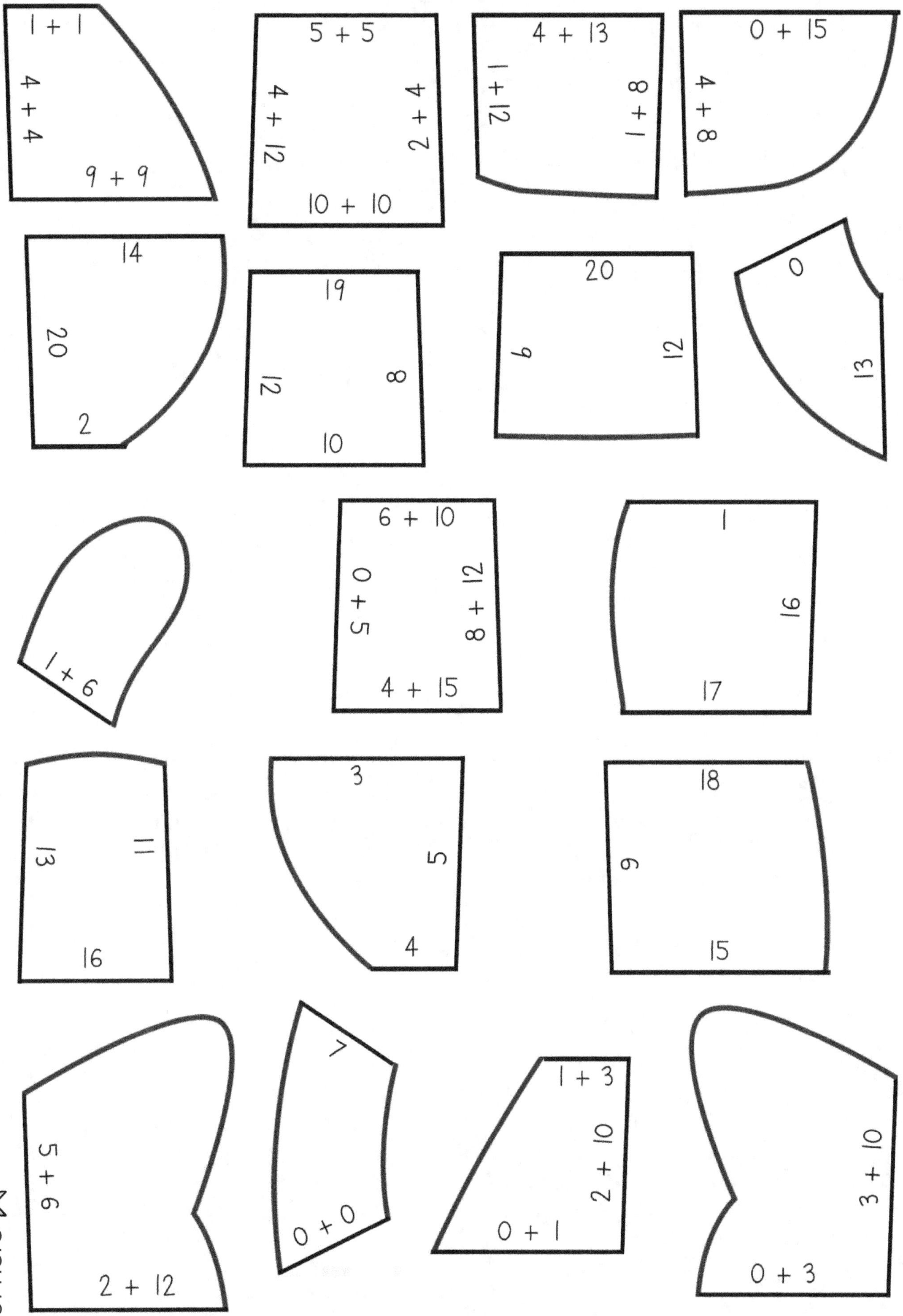
1 + 1
4 + 4
9 + 9
5 + 5
4 + 12
2 + 4
10 + 10
4 + 13
1 + 12
1 + 8
0 + 15
4 + 8
14
20
2
19
12
8
10
20
6
12
0
13
1 + 6
6 + 10
0 + 5
8 + 12
4 + 15
1
16
17
13
11
16
3
5
4
18
6
15
5 + 6
2 + 12
7
0 + 0
1 + 3
2 + 10
0 + 1
3 + 10
0 + 3

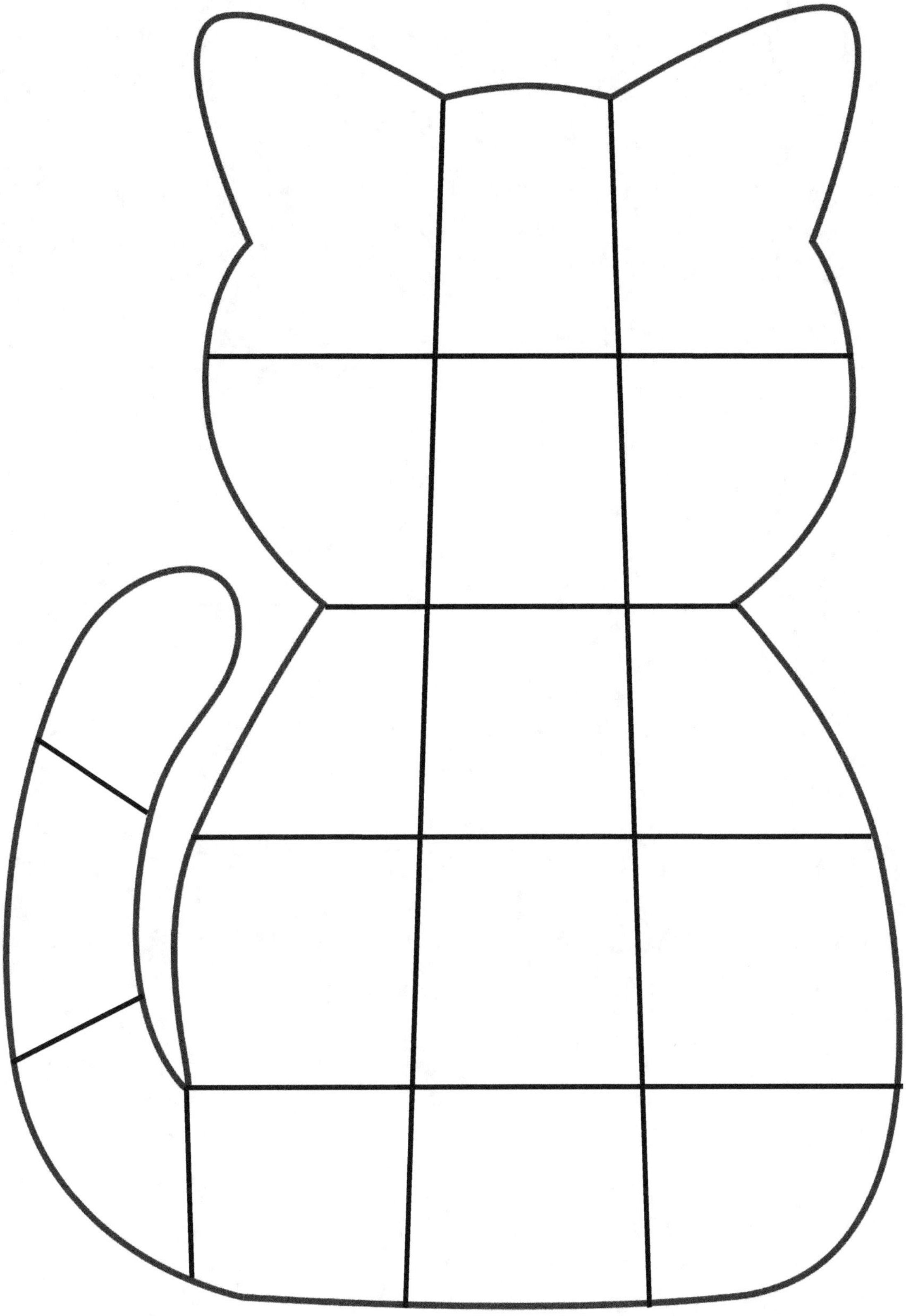

Use the puzzle mat to complete the puzzle.

Cut here ✂

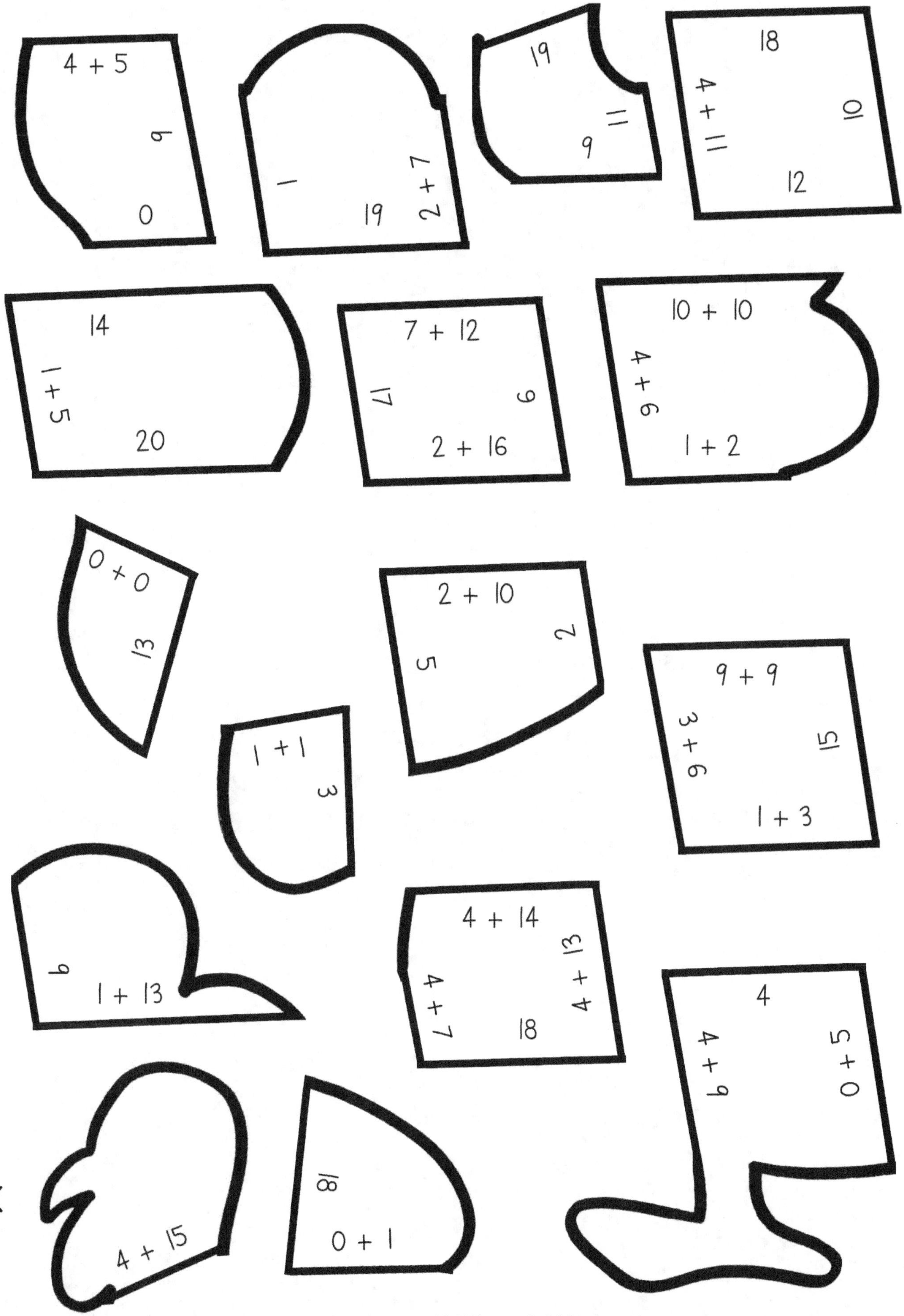
4 + 5
9
0
1
19
2 + 7
19
11
9
18
4 + 11
10
12
14
1 + 5
20
7 + 12
17
6
2 + 16
10 + 10
4 + 6
1 + 2
0 + 0
13
2 + 10
2
5
9 + 9
3 + 6
15
1 + 3
1 + 1
3
6
1 + 13
4 + 14
4 + 7
4 + 13
18
4
4 + 9
0 + 5
4 + 15
18
0 + 1

Use the puzzle mat to complete the puzzle.

Cut here ✂

12
3 + 13
7 + 8
11
2 + 9
2
2 + 4
4 + 11
10
4
8
11 + 2
1 + 6
0 + 11
5 + 7
0 + 5
14
15
0 + 15
14
13
8 + 8
9
15
6
19
4 + 14
0 + 3
12
1 + 17
1 + 3
16
18
7
10 + 10
11
4 + 9
10
4 + 15
0 + 17
13
5 + 9
20
1 + 8
9 + 11
18
16
2 + 17
0 + 8
3
4 + 6
15
1 + 1
20
17
4 + 8
2 + 12
5
19
0 + 10

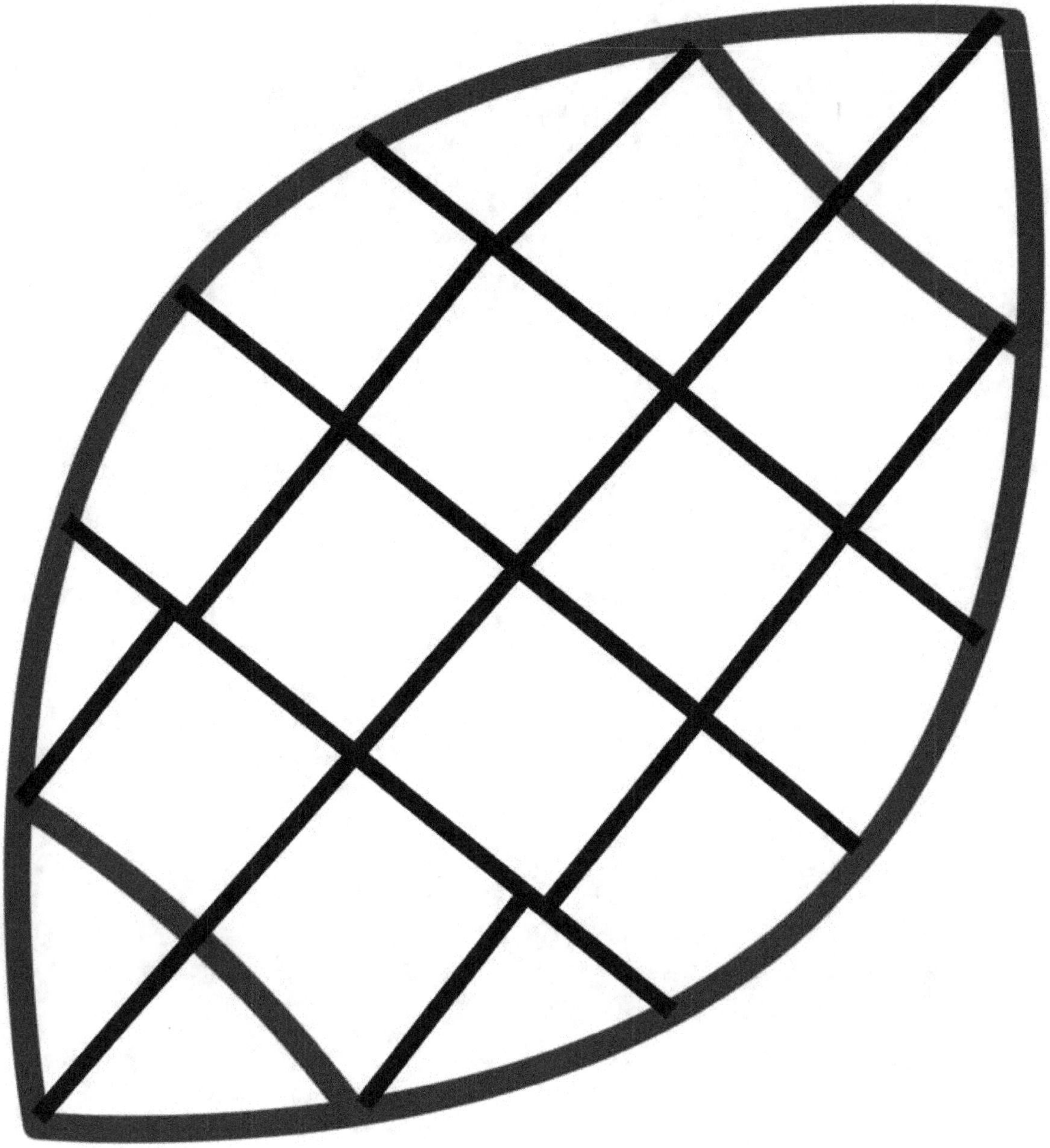

Use the puzzle mat to complete the puzzle.

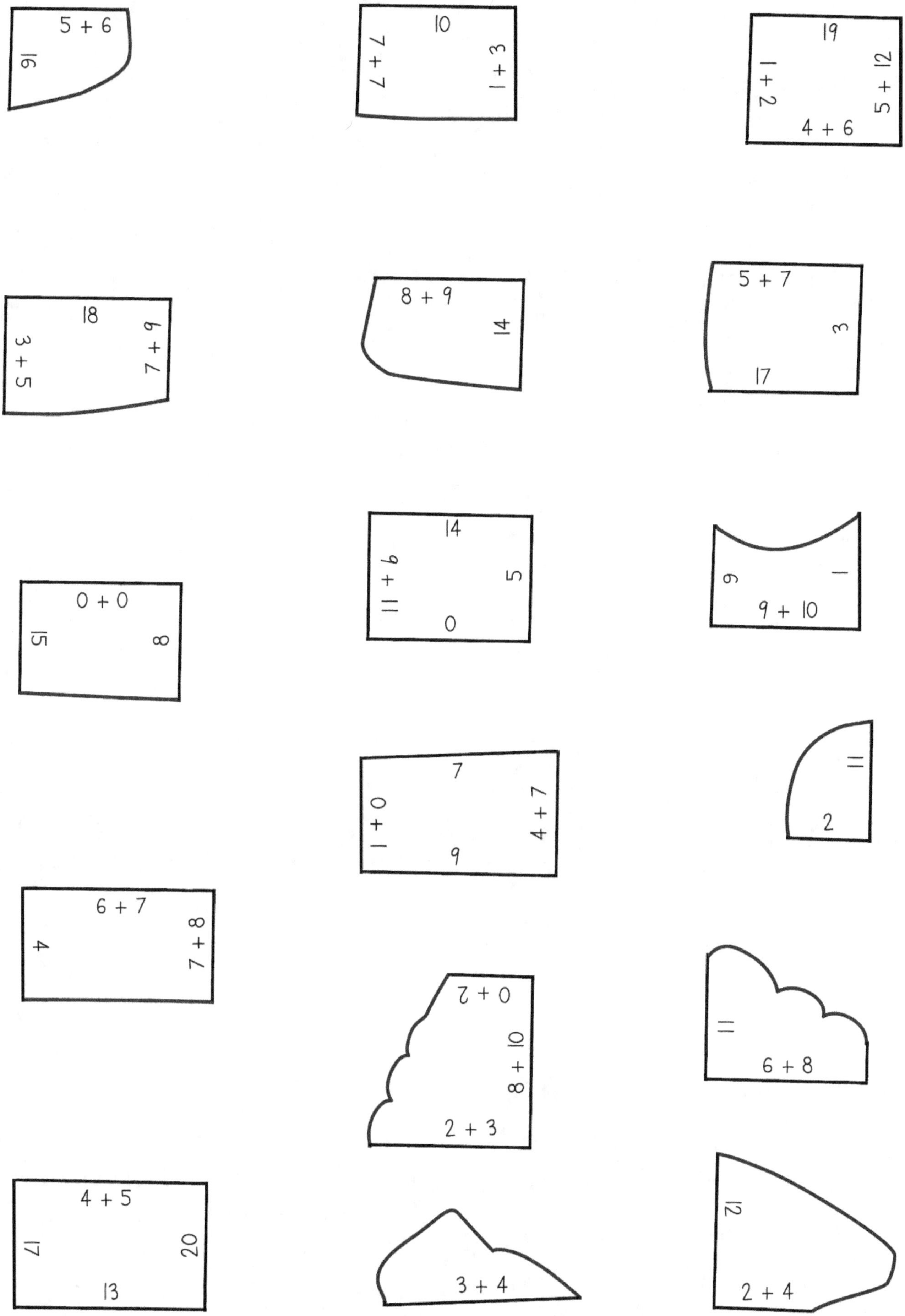
5 + 6
16
10
7 + 7
1 + 3
19
1 + 2
5 + 12
4 + 6
18
3 + 5
7 + 9
8 + 9
14
5 + 7
3
17
0 + 0
15
8
14
11 + 9
5
0
9
1
9 + 10
7
1 + 0
4 + 7
9
11
2
6 + 7
4
7 + 8
0 + 2
8 + 10
2 + 3
11
6 + 8
4 + 5
17
20
13
3 + 4
12
2 + 4

Cut here

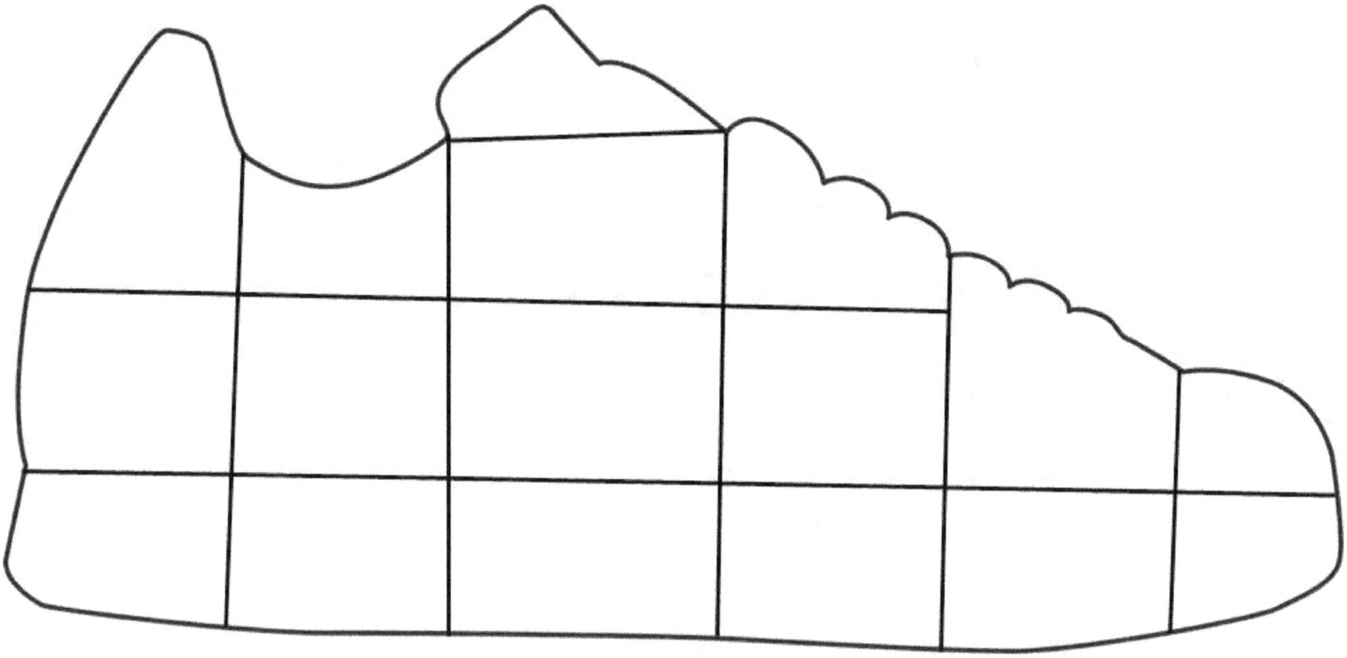

Use the puzzle mat to complete the puzzle.

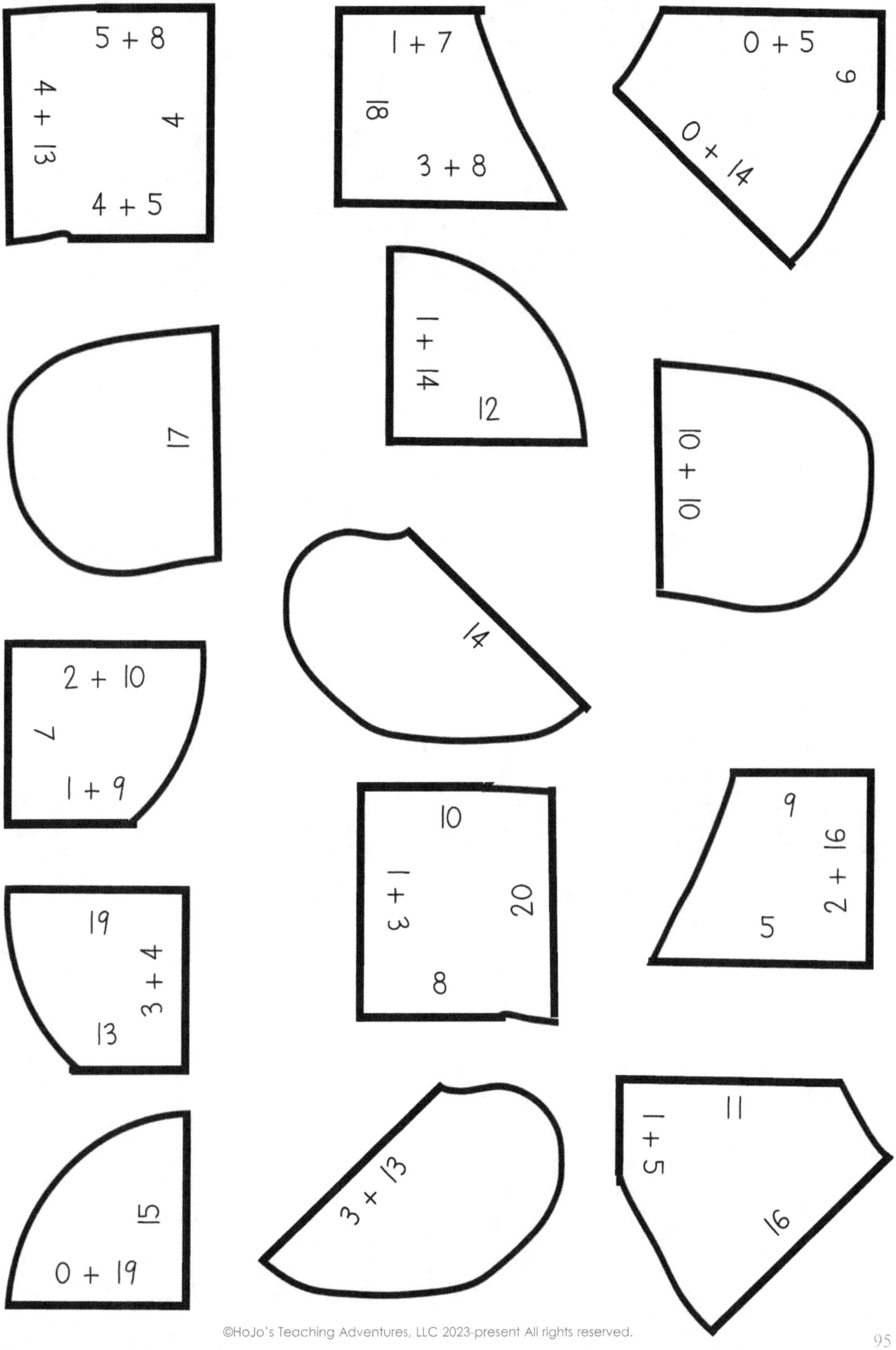
5 + 8
4 + 13
4
4 + 5
1 + 7
18
3 + 8
0 + 5
6
0 + 14
17
1 + 14
12
10 + 10
14
2 + 10
7
1 + 9
10
3 + 1
20
8
9
2 + 16
5
19
3 + 4
13
15
0 + 19
3 + 13
11
5 + 1
16

Cut here

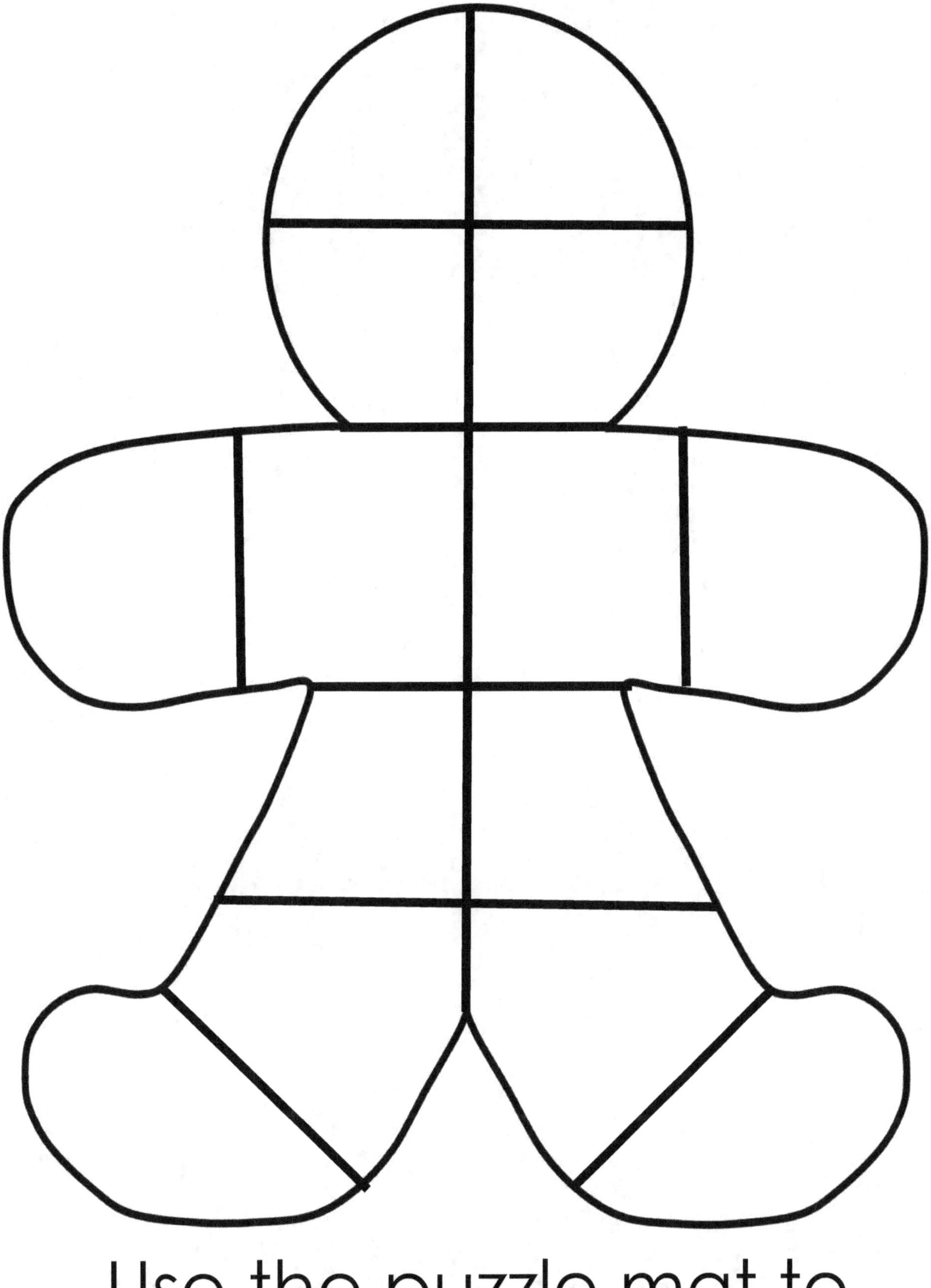

Use the puzzle mat to complete the puzzle.

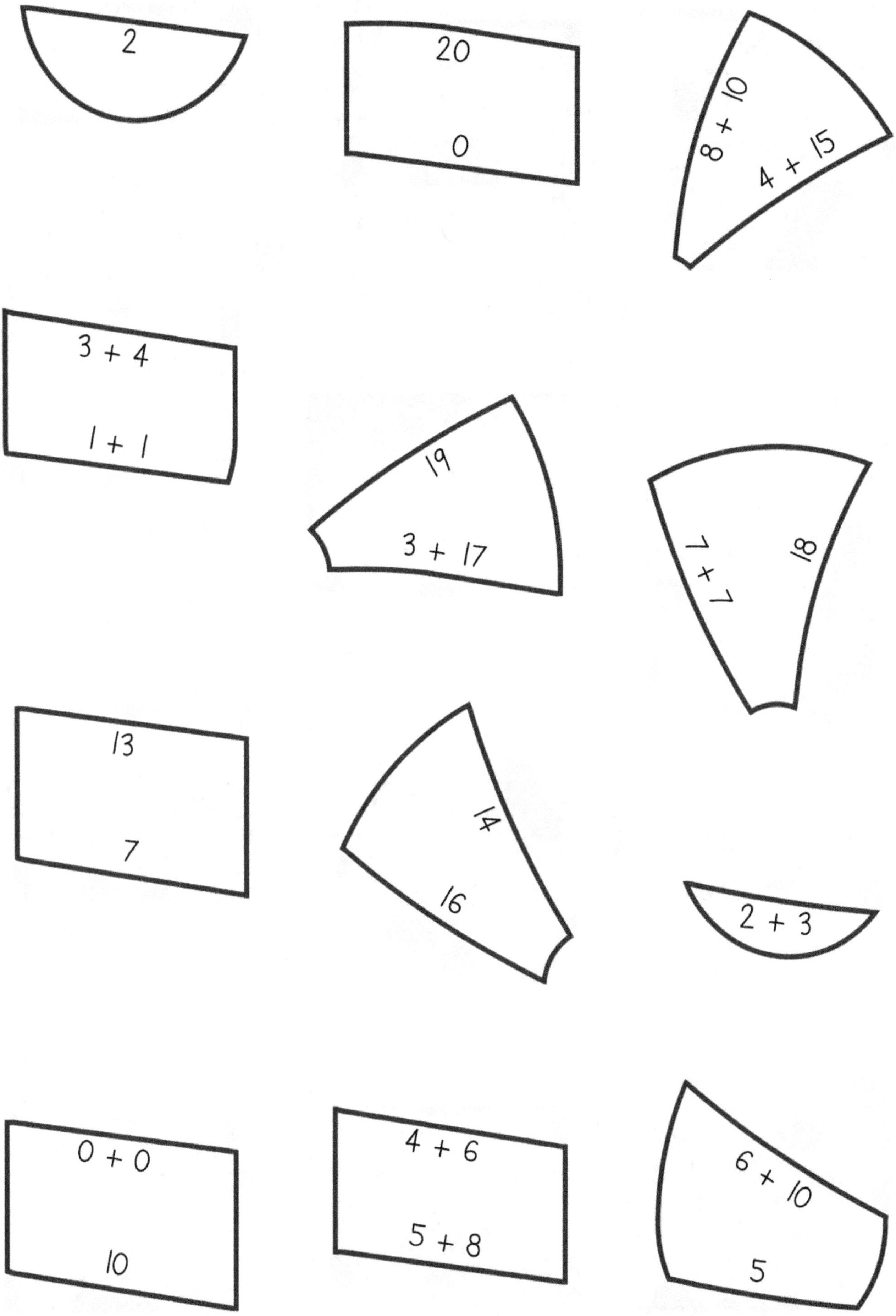

Cut here ✂

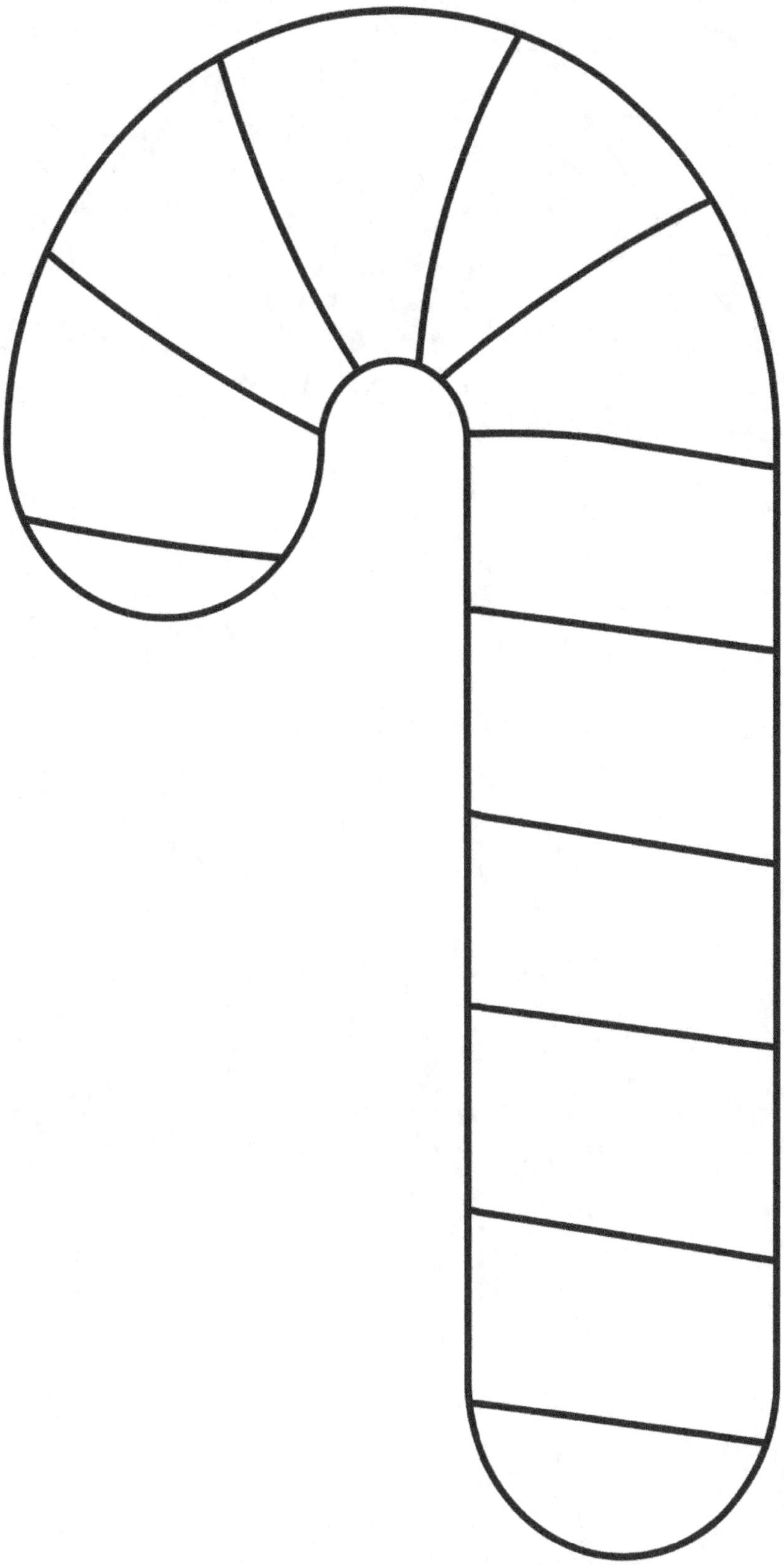

Use the puzzle mat to complete the puzzle.

Cut here

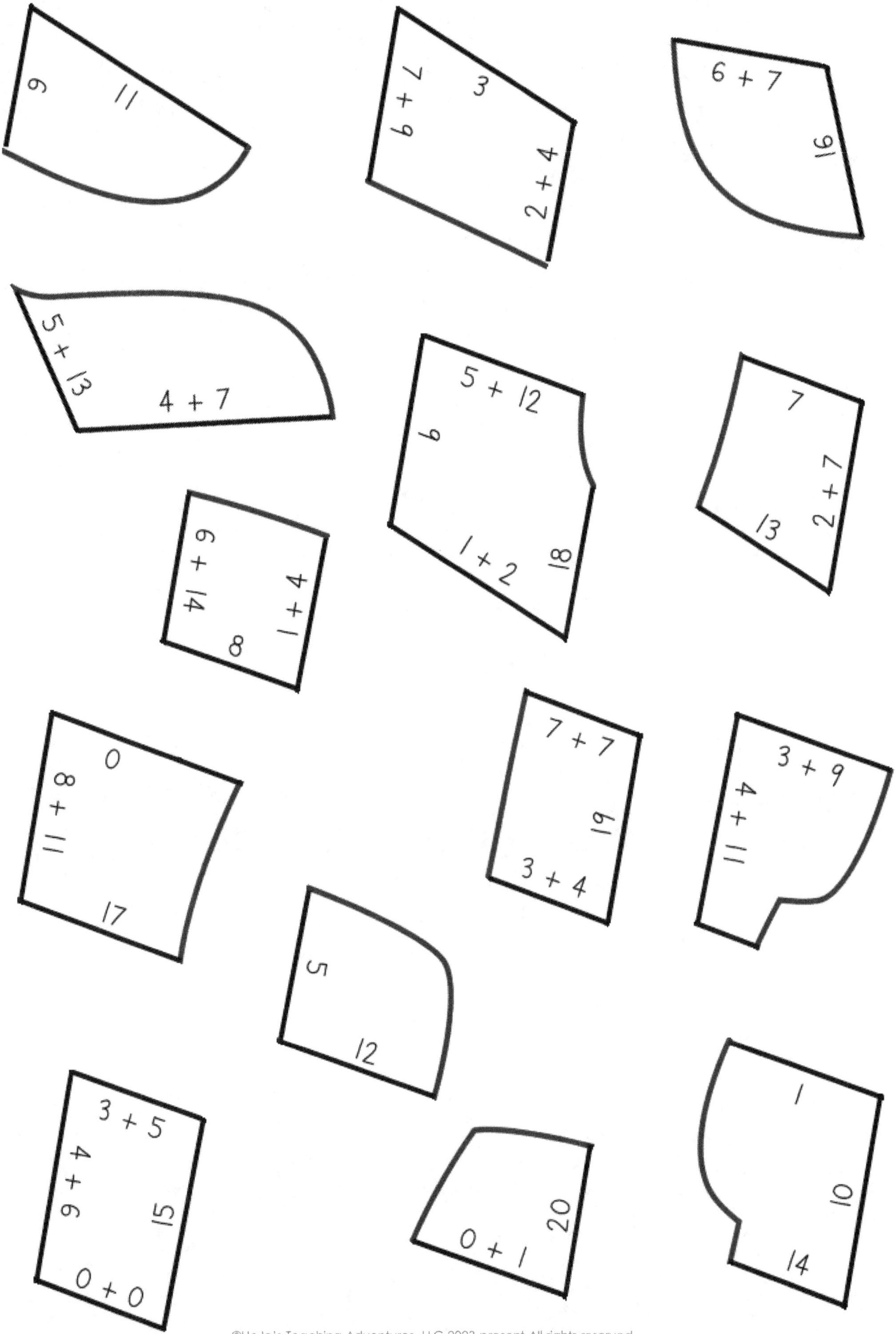
9
11
7 + 9
3
2 + 4
6 + 7
16
5 + 13
4 + 7
5 + 12
9
1 + 2
18
7
2 + 7
13
6 + 14
1 + 4
8
0
8 + 11
17
7 + 7
19
3 + 4
3 + 9
4 + 11
5
12
3 + 5
4 + 6
15
0 + 0
20
0 + 1
1
10
14

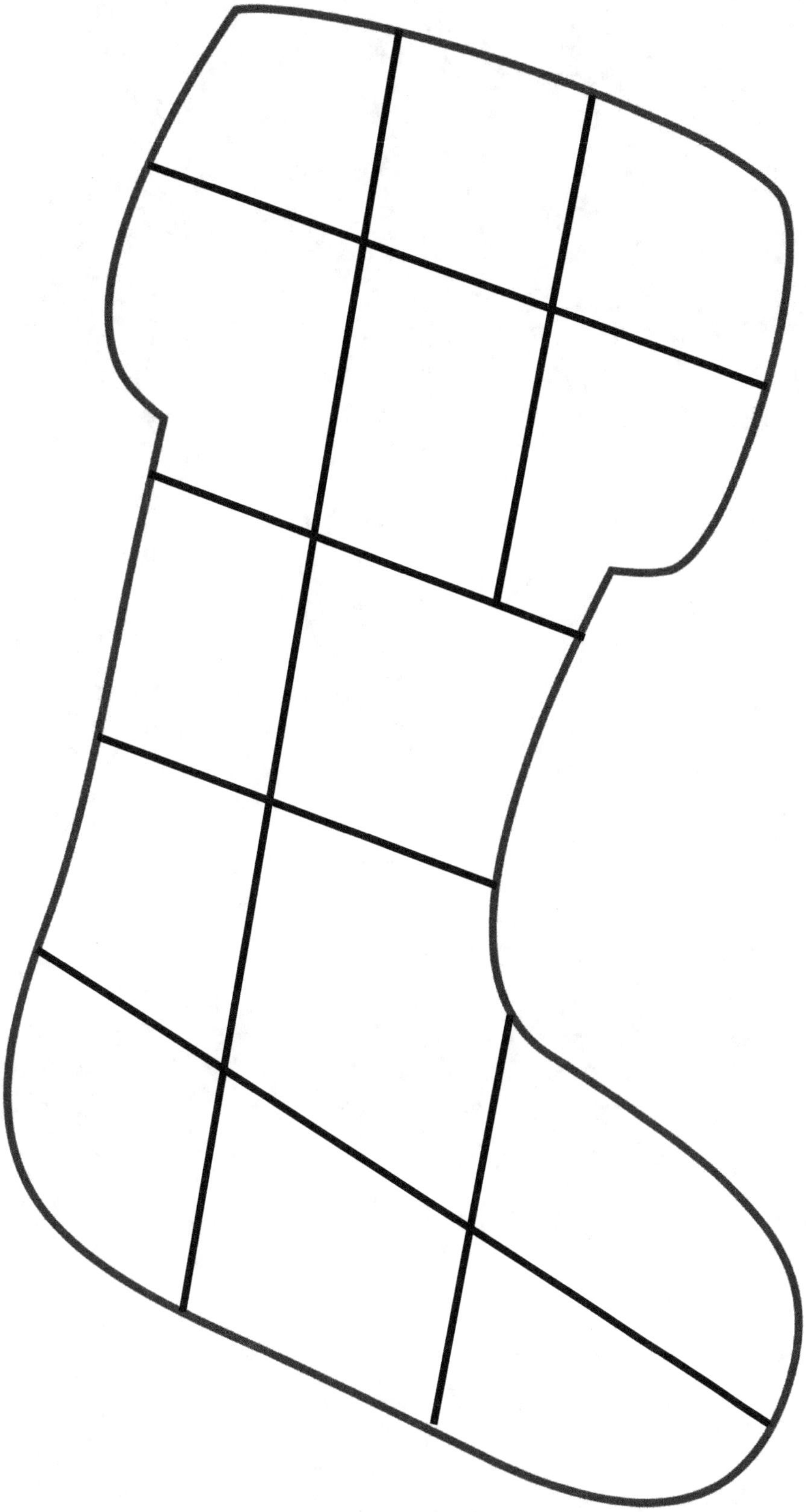

Use the puzzle mat to complete the puzzle.

Cut here

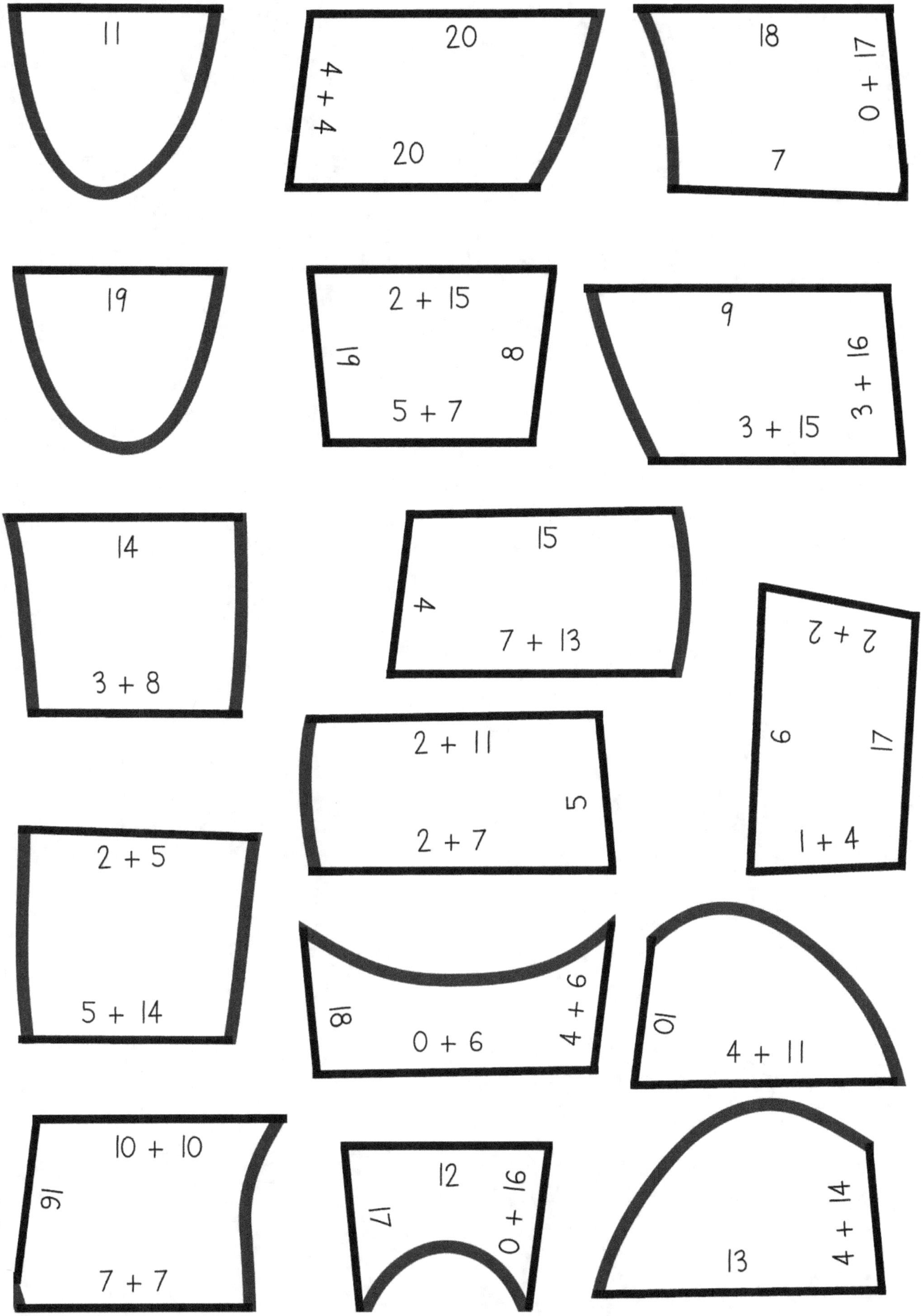
11
20
4 + 4
20
18
0 + 17
7
19
2 + 15
19
8
5 + 7
9
3 + 16
3 + 15
14
3 + 8
15
4
7 + 13
2 + 2
9
17
1 + 4
2 + 11
5
2 + 7
2 + 5
5 + 14
18
0 + 6
4 + 6
10
4 + 11
10 + 10
16
7 + 7
12
17
0 + 16
4 + 14
13

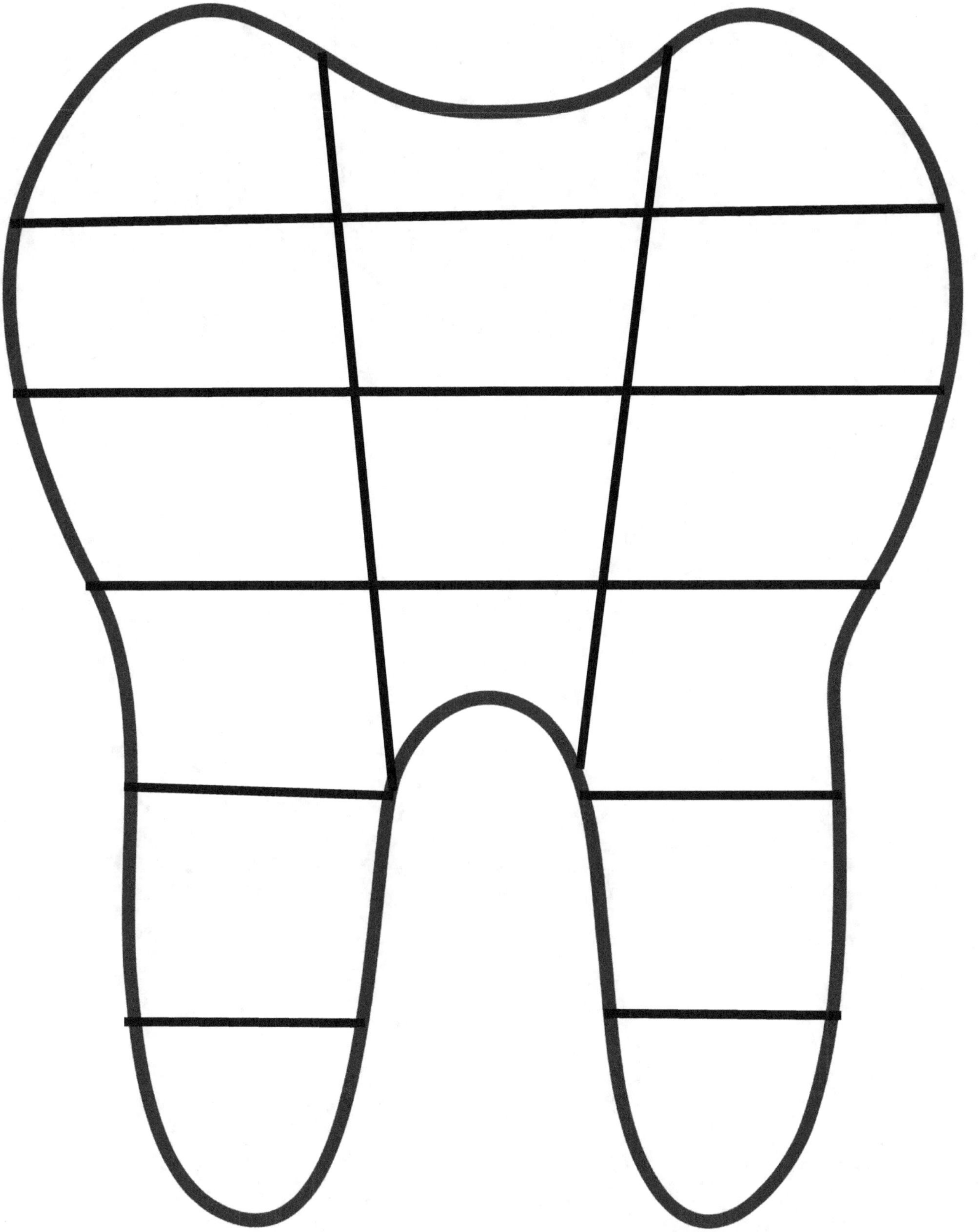

Use the puzzle mat to complete the puzzle.

Cut here

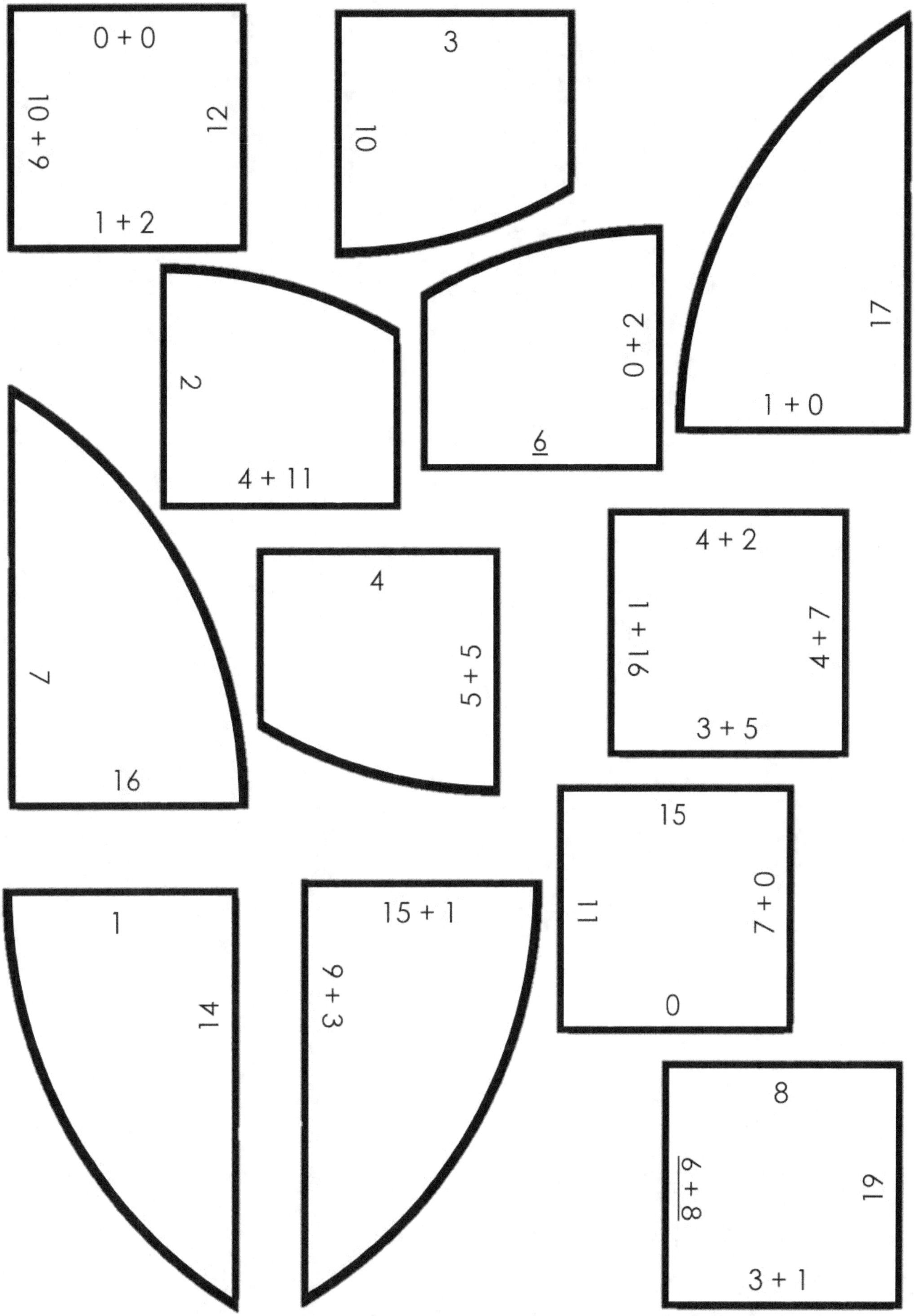
0 + 0
10 + 9
12
1 + 2
3
10
17
1 + 0
2
4 + 11
0 + 2
6
7
16
4
5 + 5
4 + 2
1 + 16
4 + 7
3 + 5
15
11
7 + 0
0
1
14
15 + 1
9 + 3
8
8 + 9
19
3 + 1

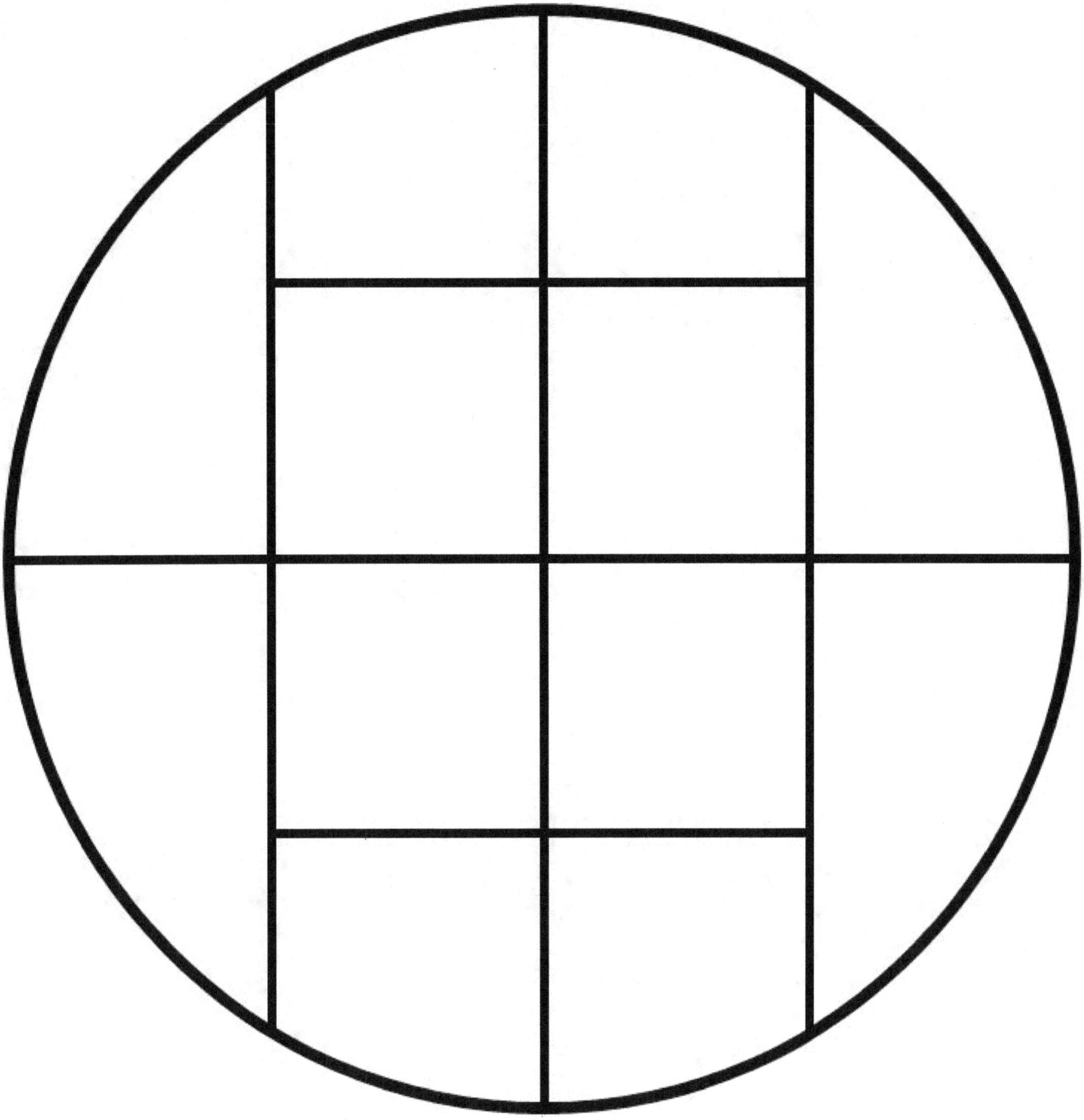

Use the puzzle mat to complete the puzzle.

1 + 3

7 + 4

18

14

3

1 + 7

11 + 4

8

4

2

4 + 5

2 + 1

5

6 + 8

15

10

3 + 2

0 + 2

8 + 2

5 + 11

3 + 15

11

16

9

Cut here

Use the puzzle mat to complete the puzzle.

Cut here ✂

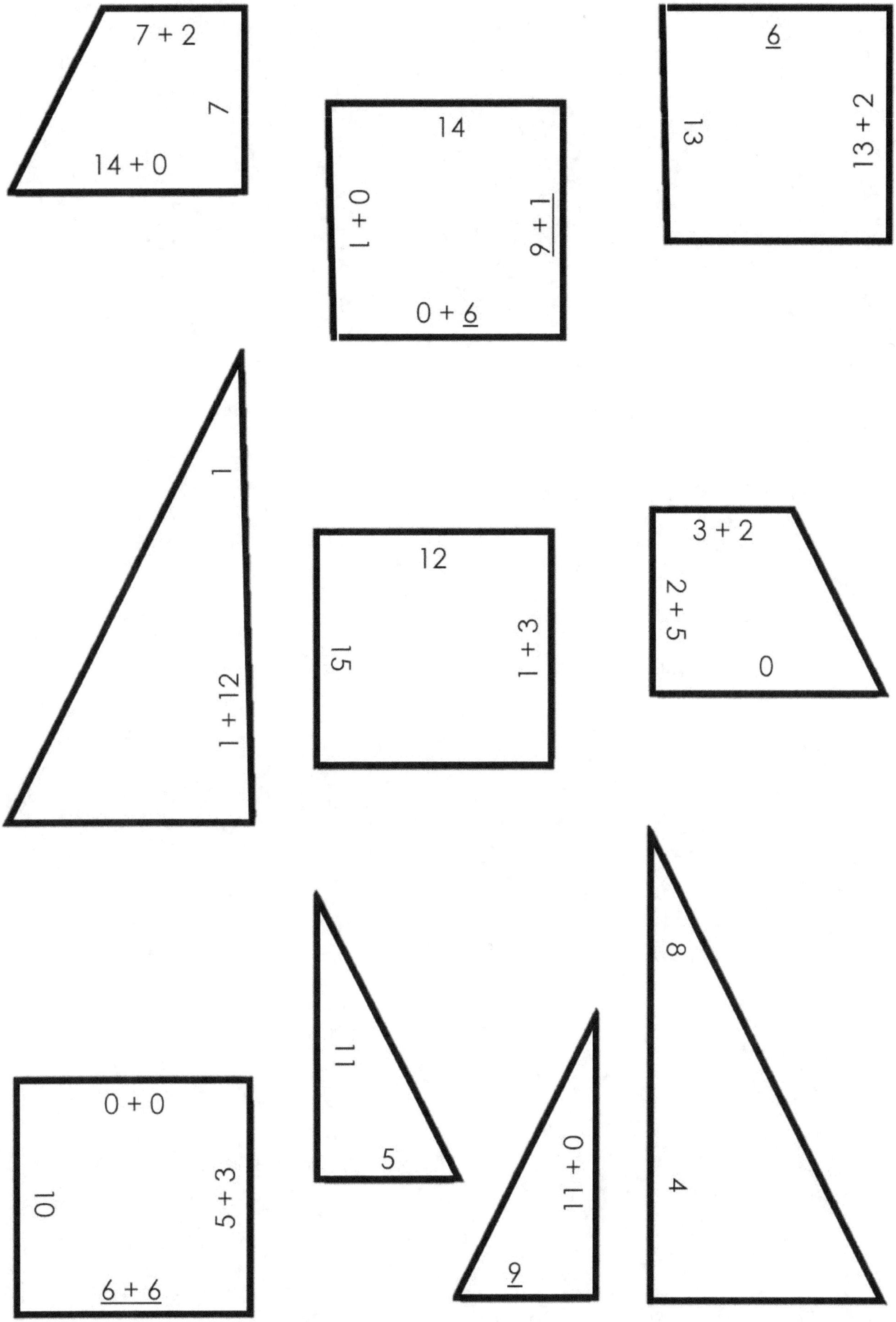
7 + 2
7
14 + 0
14
0 + 1
9 + 1
0 + 6
6
13
13 + 2
1
1 + 12
12
15
1 + 3
3 + 2
2 + 5
0
0 + 0
10
5 + 3
6 + 6
11
5
8
11 + 0
9
4

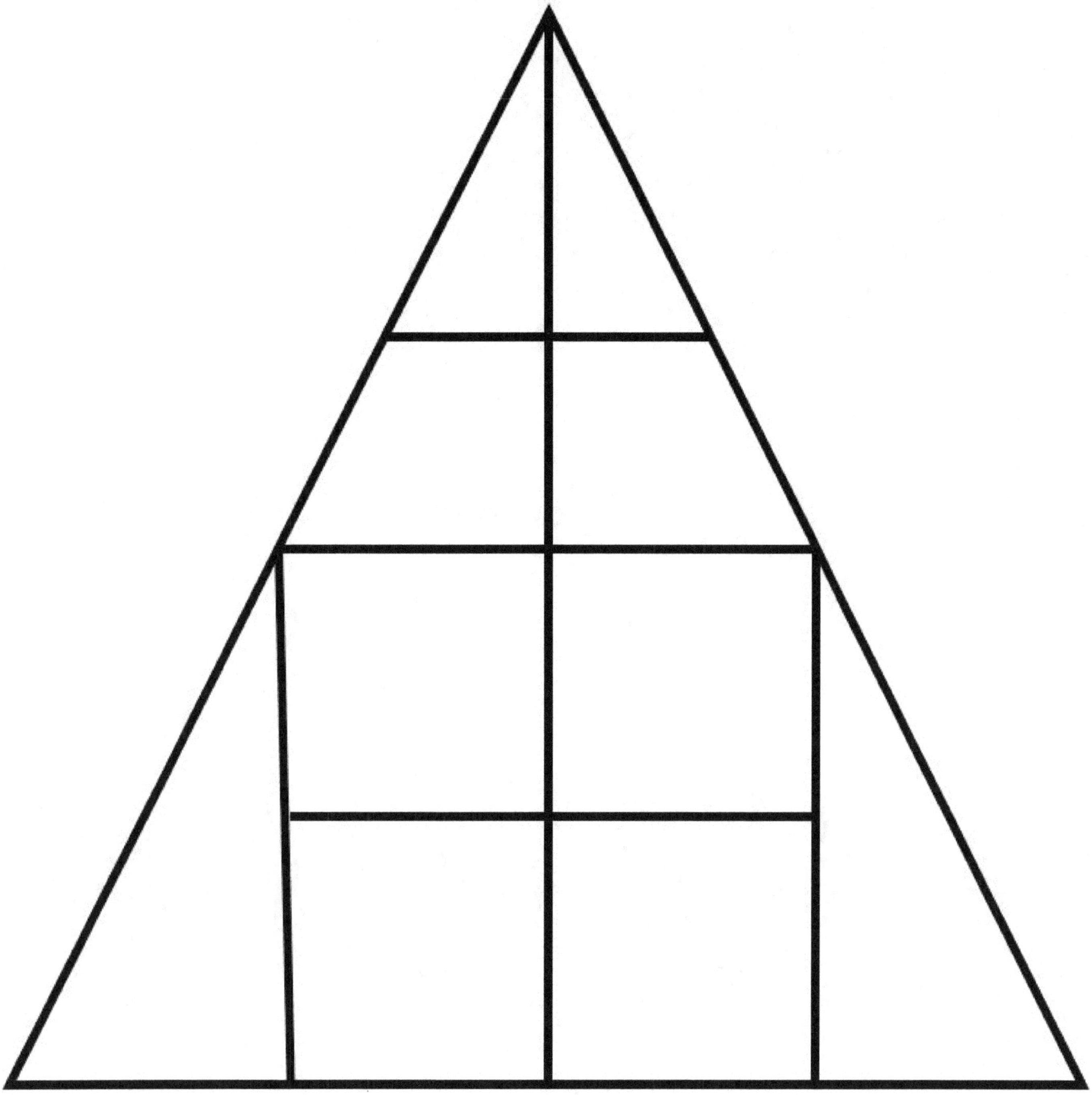

Use the puzzle mat to complete the puzzle.

Cut here ✂

Use the puzzle mat to complete the puzzle.

Hundreds Chart

1	2	3	4	5	6	7	8	9	10
11	12	13	14	15	16	17	18	19	20
21	22	23	24	25	26	27	28	29	30
31	32	33	34	35	36	37	38	39	40
41	42	43	44	45	46	47	48	49	50
51	52	53	54	55	56	57	58	59	60
61	62	63	64	65	66	67	68	69	70
71	72	73	74	75	76	77	78	79	80
81	82	83	84	85	86	87	88	89	90
91	92	93	94	95	96	97	98	99	100

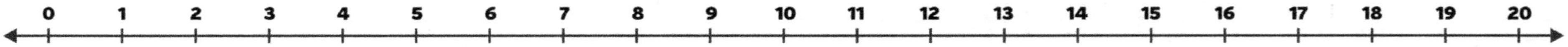
0 1 2 3 4 5 6 7 8 9 10 11 12 13 14 15 16 17 18 19 20

ANSWER KEY – Page 1

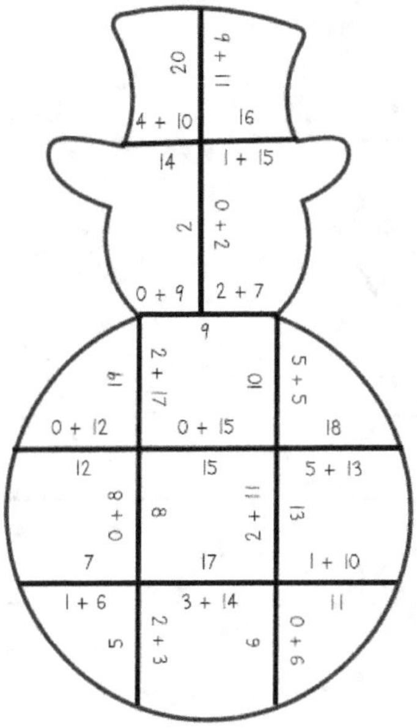

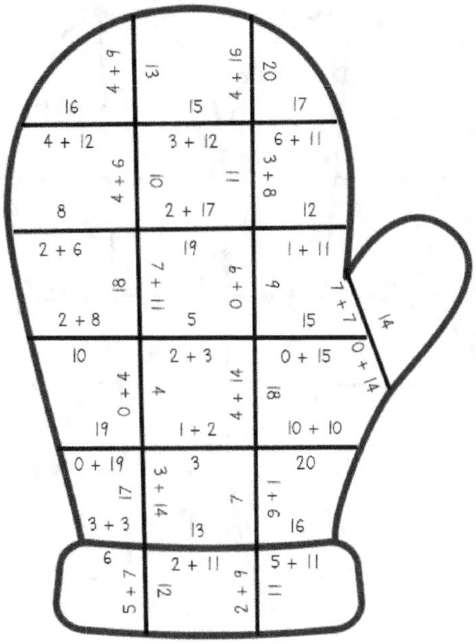

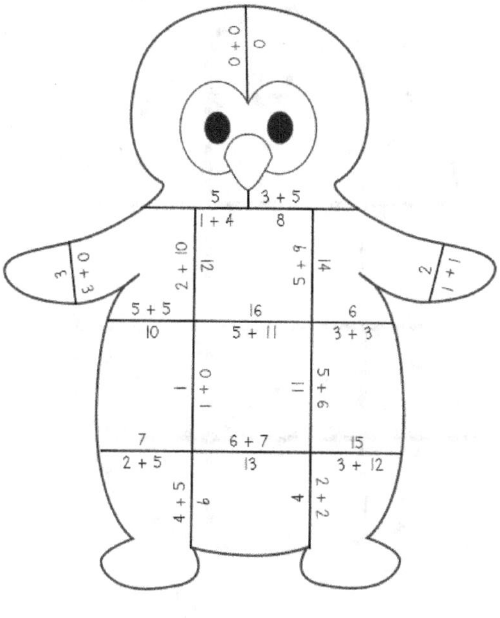

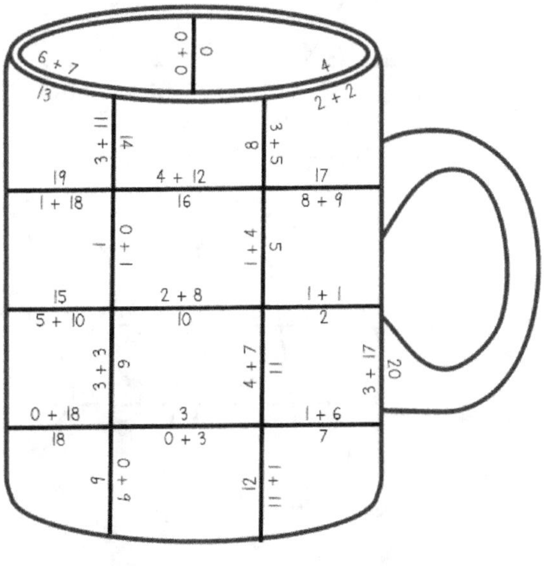

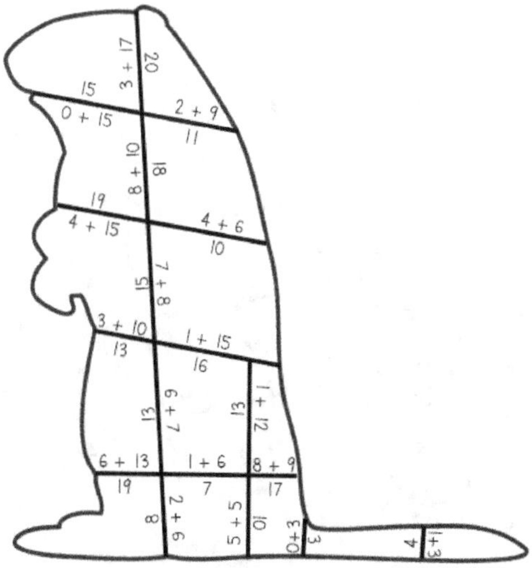

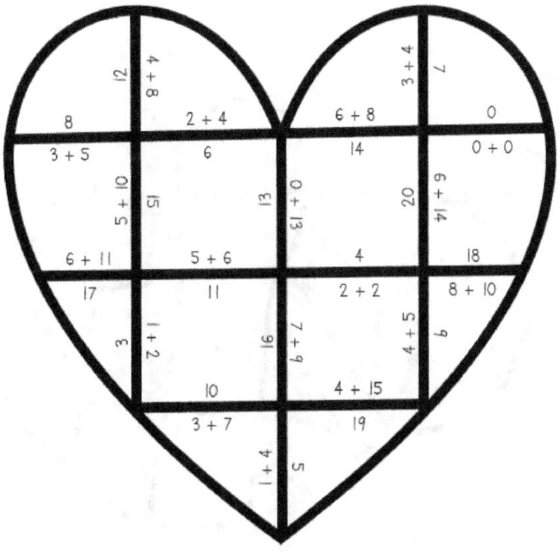

ANSWER KEY – Page 2

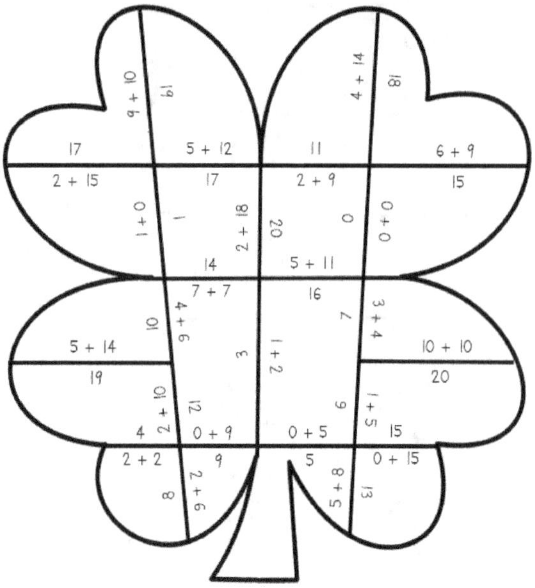

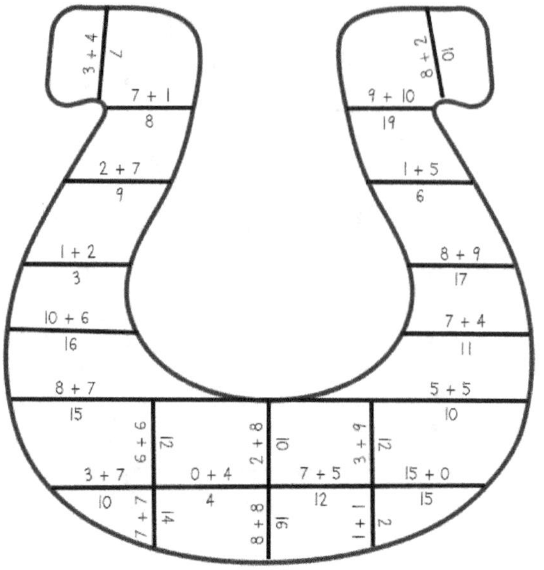

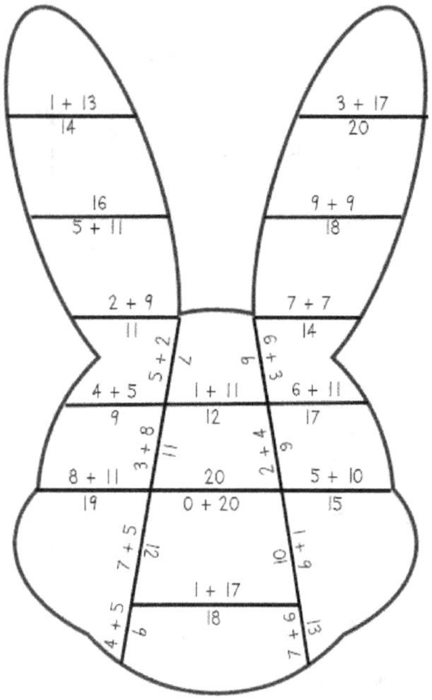

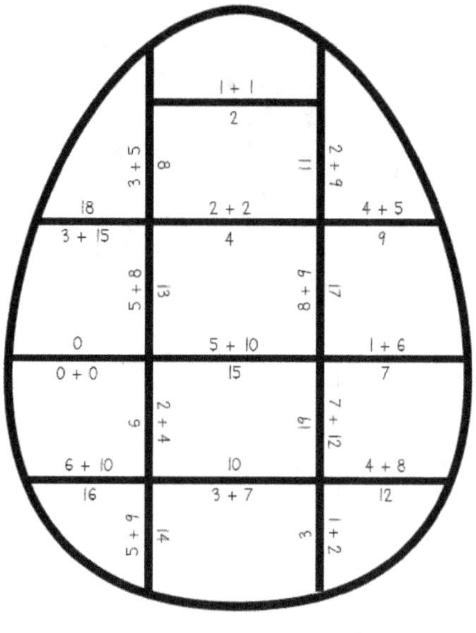

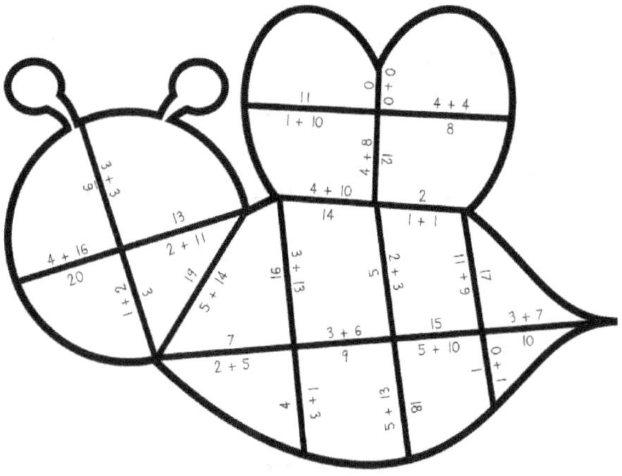

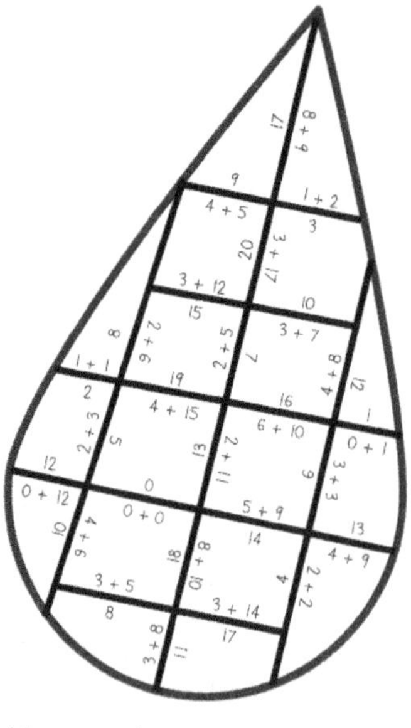

ANSWER KEY – Page 3

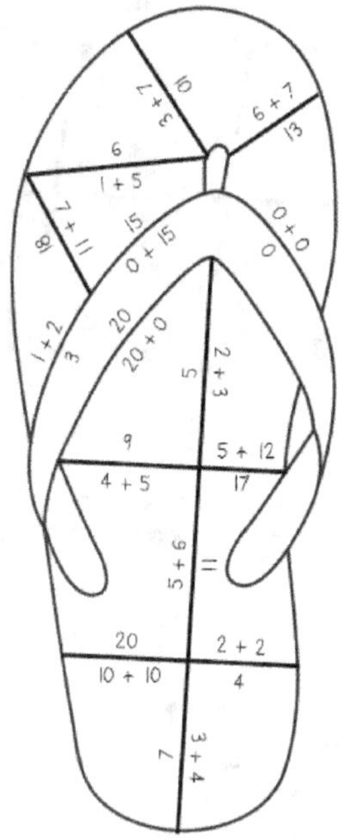

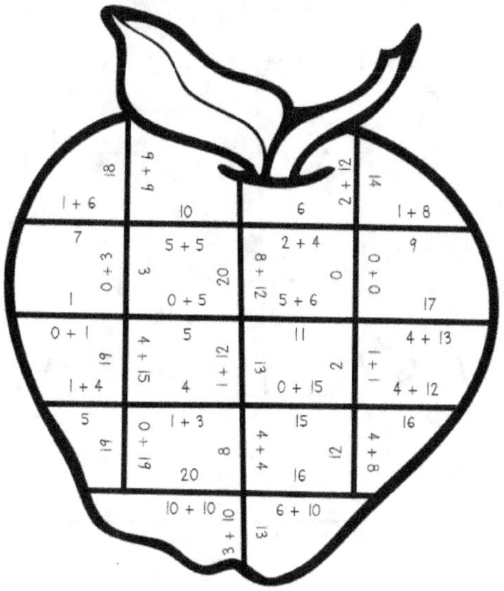

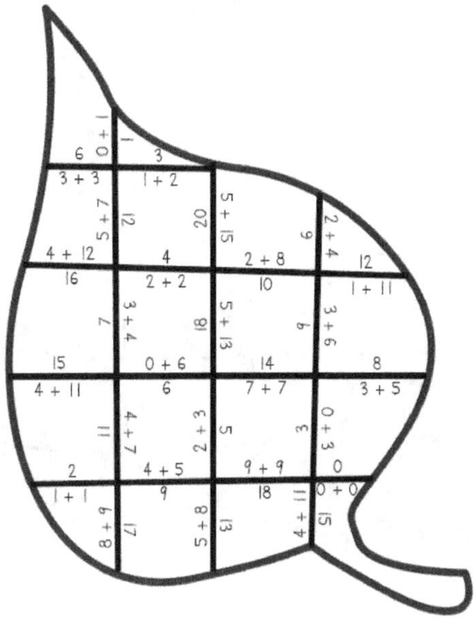

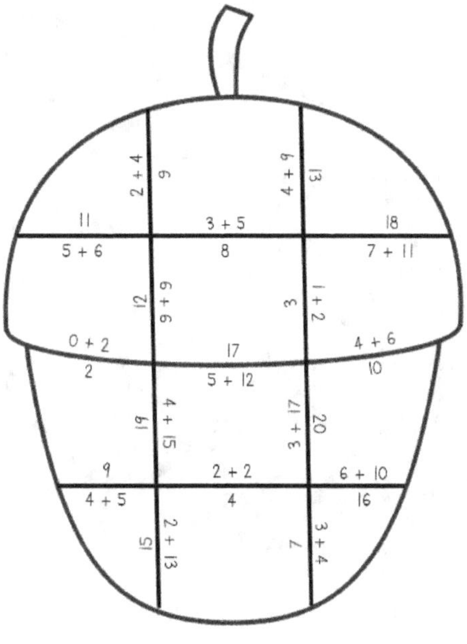

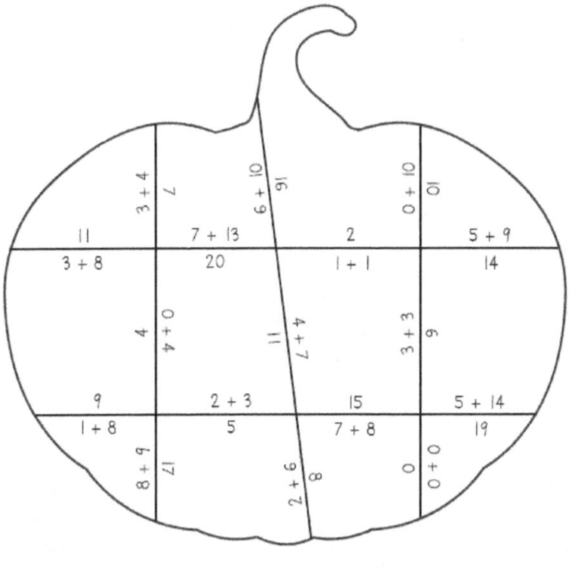

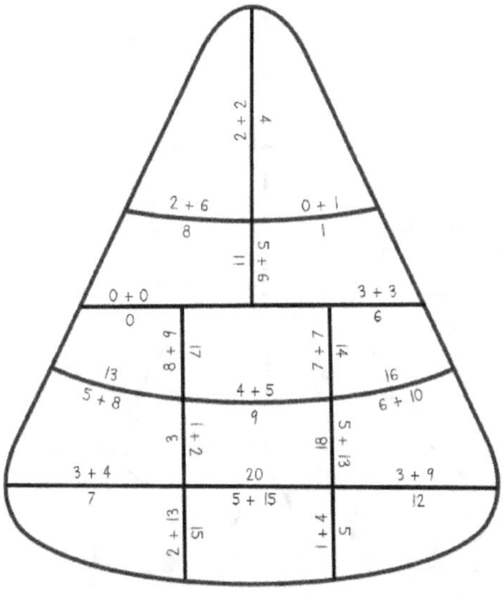

ANSWER KEY – Page 4

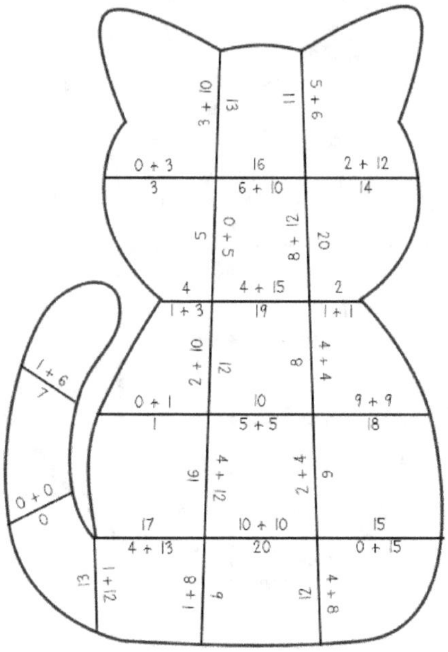

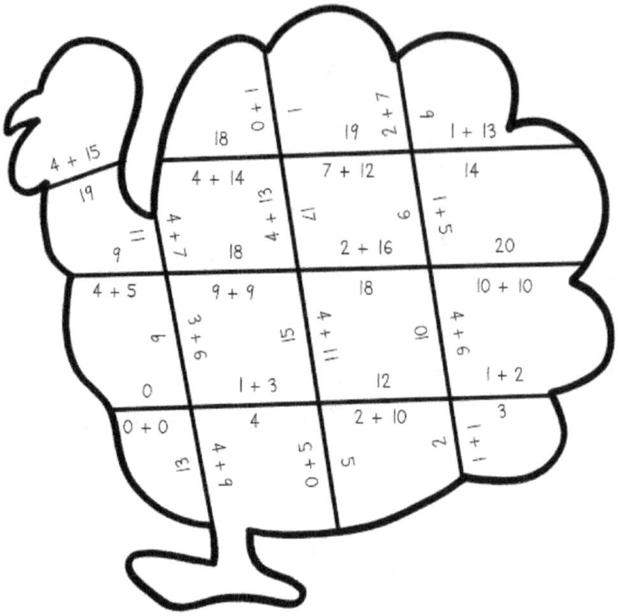

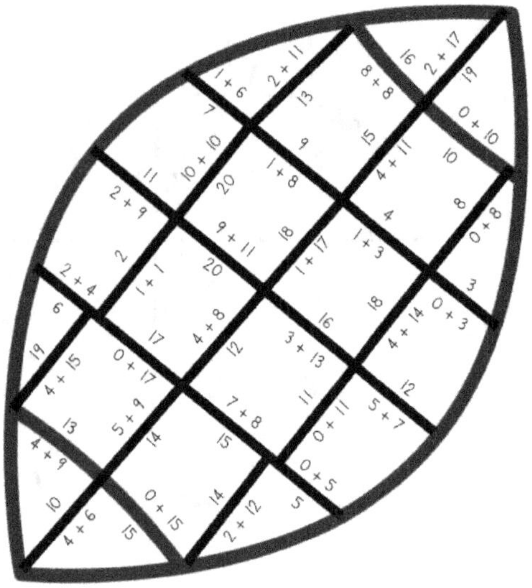

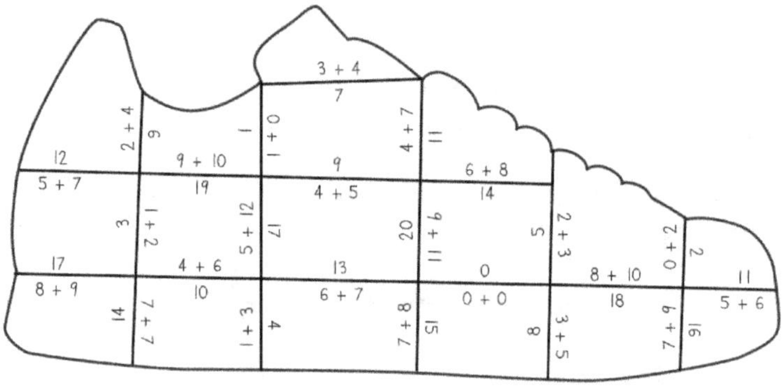

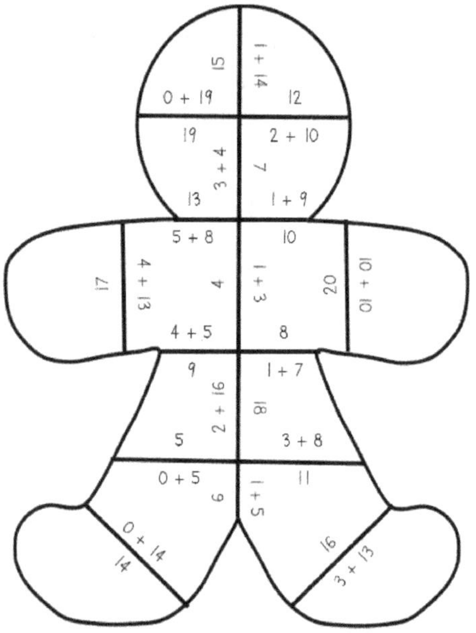

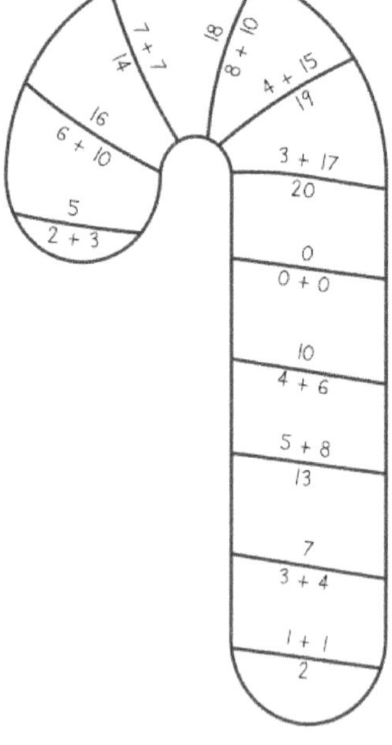

ANSWER KEY – Page 5

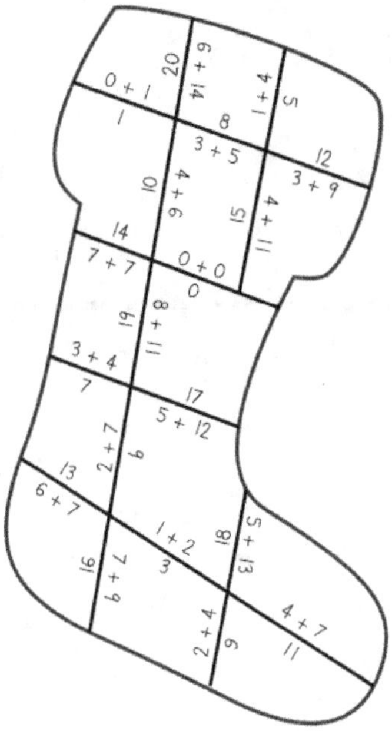

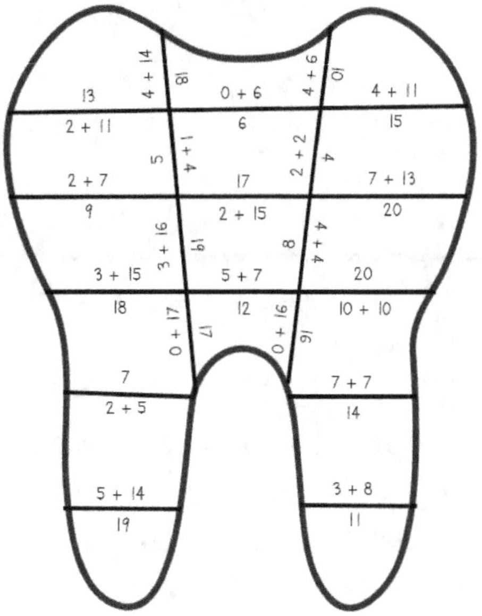

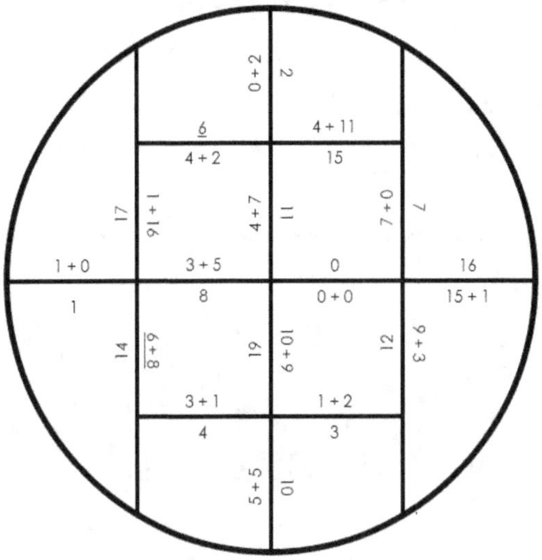

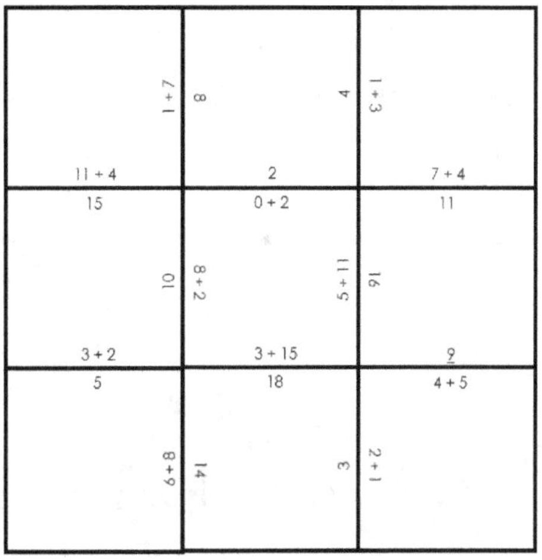

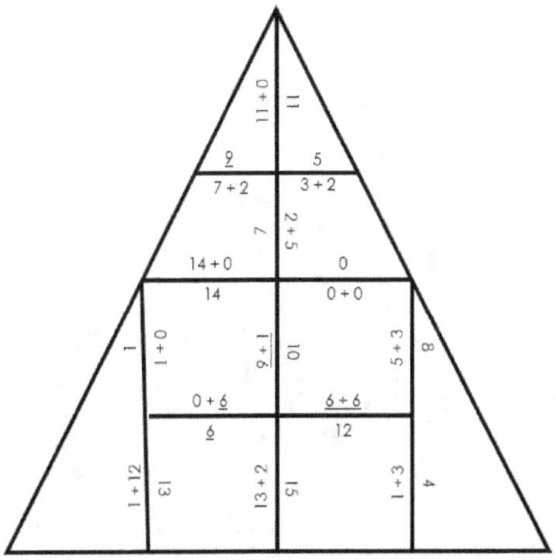

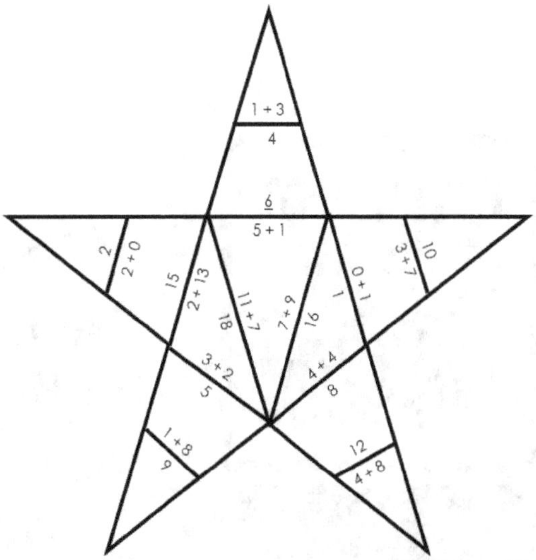

www.HoJosTeachingAdeventures.com

SCAN ME

Want FREE math puzzles you can use today?

HOJO'S TEACHING Adventures

www.ingramcontent.com/pod-product-compliance
Lightning Source LLC
Chambersburg PA
CBHW082214290526
45794CB00009B/3534